T0281147

SpringerBriefs in Applied Sciences and Technology

SpringerBriefs present concise summaries of cutting-edge research and practical applications across a wide spectrum of fields. Featuring compact volumes of 50 to 125 pages, the series covers a range of content from professional to academic.

Typical publications can be:

- A timely report of state-of-the art methods
- An introduction to or a manual for the application of mathematical or computer techniques
- A bridge between new research results, as published in journal articles
- A snapshot of a hot or emerging topic
- An in-depth case study
- A presentation of core concepts that students must understand in order to make independent contributions

SpringerBriefs are characterized by fast, global electronic dissemination, standard publishing contracts, standardized manuscript preparation and formatting guidelines, and expedited production schedules.

On the one hand, **SpringerBriefs in Applied Sciences and Technology** are devoted to the publication of fundamentals and applications within the different classical engineering disciplines as well as in interdisciplinary fields that recently emerged between these areas. On the other hand, as the boundary separating fundamental research and applied technology is more and more dissolving, this series is particularly open to trans-disciplinary topics between fundamental science and engineering.

Indexed by EI-Compendex, SCOPUS and Springerlink.

Lorrie Molent

Aircraft Fatigue Management

 Springer

Lorrie Molent
Molent Aerostructures Consulting
Melbourne, VIC, Australia

ISSN 2191-530X   ISSN 2191-5318 (electronic)
SpringerBriefs in Applied Sciences and Technology
ISBN 978-981-99-7467-2   ISBN 978-981-99-7468-9 (eBook)
https://doi.org/10.1007/978-981-99-7468-9

This Springer imprint is published by the registered company Springer Nature Singapore Pte Ltd.
The registered company address is: 152 Beach Road, #21-01/04 Gateway East, Singapore 189721, Singapore

Paper in this product is recyclable.

Acknowledgements

This work has been significantly influenced, supported and inspired by my colleagues and friends, including:

Dr. Russell Wanhill (formerly NLR, the Royal National Aerospace Centre), the Netherlands
Professor Rhys Jones AC (Monash University), Australia
Dr. Simon Barter PSM (Defence Science and Technology (DST) Group), Australia
Dr. Paul White (DST), Australia
Mr. Ben Dixon (DST), Australia
Mr. Ben Main AM (DST), Australia
Defence Aviation Safety Authority (DASA), Australia
Group Captain R. Singh DASA RAAF
Capability Acquisition and Sustainment Group (CASG), Australia
Dr. Joseph P. Gallagher (formerly USAF), USA
Mr. Don Polakovics (formerly US NAVAIR), USA
Mr. Rick Ryan (formerly US NAVAIR), USA
Professor David Hoeppner (formerly University of Utah), USA

Contents

About the Author

Mr. Lorrie Molent is an acknowledged authority in the fields of aircraft structural integrity, structural mechanics, structural and fatigue testing, advanced composite bonded repair, aircraft vulnerability, and aircraft accident investigation. He has more than 280 publications in these technical areas. Mr. Molent is a qualified aircraft accident investigator. He has been attached to both the then Australian Civil Aviation Department and the US Navy NAVAIR Structures in Washington D.C. as an airworthiness engineer and has a broad network of collaborators nationally and internationally. Until October 2020, Mr. Molent was the Australia's Defence Science and Technology Group's Head of Emerging Aircraft Structural Integrity. In 2010, he was presented with the Minister's Award for Achievements in Defence Science, and he was awarded with a member of the Order of Australia (AM) in 2016. He also has numerous other team achievement awards. Currently, he works as a consultant and trainer in aircraft structural integrity, air accident investigation, and novel fatigue analyses methods.

Chapter 1
Introduction

There are many ways an airframe can structurally fail, fatigue is one. Fatigue occurs under cyclic loading and can significantly degrade the operational capability and safety of metallic aircraft components and structures. A primary requirement for the design of modern aircraft structures is establishment of the safety and durability of the airframes under the anticipated service loads. Design lives are normally estimated via fatigue design analyses and certification testing, and much effort is spent in establishing the certified fatigue lives for airframes. However, it is often found that aircraft usages and configurations vary considerably from the design assumptions. This happens owing to variations and changes in mission types and durations, structural weight increases and—in the case of military aircraft—changes in the external stores configurations.

While available design standards (e.g. [1, 2]) provide guidance on good fatigue design principles and—to a lesser extent—through-life management principles, they are often insufficient when a fleet of aircraft operate, even partly, outside the design usage assumptions. Therefore, additional and even alternative means are required to ensure continued airworthiness under the actual operational conditions.

This book is adapted and extended from [3] and presents an overview of a through-life fatigue management philosophy aimed and used to ensure the safe and continued operation of metallic aircraft structures. This includes periods of extended life that are almost inevitably required when the original design lives are approached. The book summarizes new approaches to life assessment as well as recent and new observations of the behaviour of fatigue cracks in actual structures under realistic loading conditions. More specifically, it considers concisely the following aspects:

- Existing design standards.
- The importance and significance of determining and knowing the operational loads applied to individual aircraft.
- The importance of Full-Scale airframe Fatigue Test(ing), FSFT (or Full-Scale Durability Test(ing), FSDT).
- Causes of fatigue crack nucleation and growth in production aircraft structures.

© The Author(s), under exclusive license to Springer Nature Singapore Pte Ltd. 2024
L. Molent, *Aircraft Fatigue Management*, SpringerBriefs in Applied Sciences
and Technology, https://doi.org/10.1007/978-981-99-7468-9_1

- Behaviour of the critical cracks that could lead to fatigue failure.
- New/modified methods and criteria for the interpretation of FSFTs and in-service cracking.
- Probability of failure (POF) and scatter in fatigue lives.
- Some repair and life-extension techniques.

An overview of conventional aircraft structural integrity management and the associated tools is not included, since these are well covered in the literature (e.g. [4–8]). Fracture mechanics is generally accepted as essential to the design of metallic aircraft structures, notably for predicting and validating the fatigue crack growth (FCG) and residual strength properties. Applications of Linear Elastic Fracture Mechanics (LEFM), employing the stress intensity factor K and its derivative parameters, began in the early 1970s [6], and would be expected by now to have reached a mature status. Hence, one might also expect that LEFM methods can often be used for analysing service fatigue failures. However, much practical experience by the author and colleagues, garnered over a period of about 30 years, has shown that only non-LEFM analyses are *directly* useful for determining service failure FCG behaviour. This does not exclude the pragmatic use [9] of LEFM parameters, which can assist in analysing FCG in full-scale, component and specimen tests, and such analyses can be used—within limits—to predict some aspects of the FCG behaviour of in-service structures and components. An important shortcoming is the difficulty that LEFM modelling has in describing short crack growth, which may constitute a significant part of the FCG life, as will be shown in this book.

A broad schematic of the various LEFM and non-LEFM FCG modelling processes is shown in Fig. 1.1. On the left-hand side are the 'traditional' models [6]. These have a pedigree reaching back to the early 1970s, as mentioned earlier, and have been much used in life predictions for military aircraft and some civil aircraft. The right-hand side shows the models that have been developed from extensive observations of exponential FCG. The basic difference between these types of models is that the non-LEFM models have been derived from short-to-long FCG, removing problems experienced with LEFM models in the short crack regime.

The non-LEFM models are discussed in Chaps. 4 and 5 of this book. In addition, the reader is directed to Refs. [9, 10] for more details and examples of some of the non-LEFM analytical methods. Also, an essentially up-to-date history of airframe FSFT in Australia has been presented by the author in [11].

The contents of this book have been developed through the pragmatic need to safely manage in-service aircraft structures for extended periods and with limited resources. Owing to the subject of this book being predominantly sponsored by the Royal Australian Air Force (RAAF), many of the readily available peer-reviewed references are necessarily those authored or co-authored by the author.

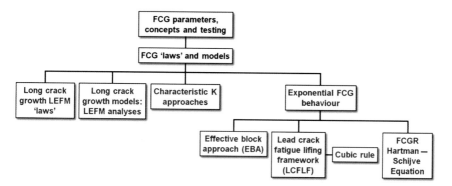

Fig. 1.1 Schematic of LEFM and non-LEFM (exponential) FCG models: courtesy of R. J. H. Wanhill

References

1. Aircraft Structures, Joint Service Specification Guide JSSG-2006. US Department of Defense (1998)
2. Ministry of Defence, Defence Standard 00-970 Issue 1, Design and Airworthiness Requirements for Service Aircraft, Volume 1-Aeroplanes, Amendment 13, December (1994)
3. L. Molent, General fatigue lifing of metallic aircraft structures, DSTG-RR-0467, Defence Science and Technology Organisation, Melbourne, Australia (2020)
4. S. Suresh, *Fatigue of Materials* (Cambridge University Press, UK, 1998)
5. A.F. Grandt Jr., *Fundamentals of Structural Integrity: Damage Tolerant Design and Nondestructive Evaluation* (Wiley, USA, 2004)
6. R.J.H. Wanhill, S. Barter, L. Molent, Fracture mechanics in aircraft failure analysis: uses and limitations. Eng. Fail. Anal. **35**, 33–45 (2013)
7. MIL-STD-1530D (W/ Change-1), Department of Defense Standard Practice: Aircraft Structural Integrity Program (ASIP), October (2016)
8. Fracture control implementation handbook for payloads, experiments, and similar hardware, NASA-HDBK-5010, w/change 1:revalidated 2017, USA (2017)
9. R.J.H. Wanhill, S. Barter, L. Molent, *Fatigue Crack Growth and Lifing Analyses for Metallic Aircraft Structures and Components*. Springer Briefs in Applied Science and Technology The Netherlands, (2019)
10. L. Molent, B. Dixon (Review Paper), Airframe metal fatigue revisited. Int. J. Fatigue **131** (2020)
11. L. Molent, The history of structural fatigue testing at Fishermans Bend Australia, DSTO-TR-1773, Defence Science and Technology Organisation, Melbourne, Australia, October (2005)

Chapter 2
Fatigue Demystified

The study of metal fatigue dates back to 1844 [1]; the first meeting related to aircraft fatigue was held in Melbourne in 1946 [2]. Despite many breakthroughs in understanding, there is still much debate over the significance of some actual and potential influences, and the ability to accurately predict fatigue lives. Also, unanticipated fatigue failures still occur [3, 4]. Some known and debated aspects include

- Fatigue is the degradation of a material under cyclic loading: ✓ (i.e. no debate).
- Fatigue is a process whereby cracking occurs under the influence of repeated or cyclic stresses, which are normally substantially below the nominal yield strength of the material: ✓.
- The only quantifiable measure of fatigue is crack growth: ✓. However, this is debated in the open literature by continuum mechanics enthusiasts. By definition, they do not countenance the presence or occurrence of material discontinuities and cracks.
- Life is related to stress magnitude and to a lesser extent the mean stress: ✓.
- Life is influenced by the environment and ambient temperature: ✓.
- Fatigue crack growth (FCG) results in so-called striations in some materials. One cycle often relates to one striation, but not in vacuo or at low ΔK values; nor necessarily under variable amplitude (VA) loading: ✓/X. This topic is still debated with respect to different materials.
- Fatigue nucleation is mainly a surface-influenced phenomenon: ✓. But there is much debate directed to some specific situations. An uncontroversial example is shot peening, which can result in internal nucleation and longer life:✓.
- There is a cycle frequency effect (low frequency tends to produce broader striations or even change the fracture mode): ✓. An association with environmental influences has been observed for some materials, including aerospace aluminium alloys.
- FCG rate (FCGR) is proportional to stress ratio ($R = \sigma_{min}/\sigma_{max}$): ✓/X. But not for short cracks or some steels.

- Overloads lead to FCG retardation: $\sqrt{}/$**X**. There is actually a short acceleration followed by predominantly retardation, depending on the overload magnitude.
- The stress intensity factors K, ΔK or Kmax can be used as unifying similitude parameters: $\sqrt{}/$**X**. There is still some debate and uncertainty, but the usefulness of these parameters is undeniable for constant amplitude (CA) loading, less so for VA loading.
- Crack advance is entirely by (alternate) slip or else combined with 'decohesion': $\sqrt{}/$**X**. There is still lively debate among micromechanism specialists, and the influence of environment must also be considered, although often ignored.
- Initiation/nucleation period is insignificant: $\sqrt{}$. Some debate, particularly among aeroengine designers, and mainly about engineering 'conveniences' rather than physical realities.
- There is a cyclic ΔK threshold (ΔK_{THRS}): **X**. Thresholds appear not to exist, or are very low, for physically short cracks or naturally short-to-long crack growth in aluminium alloys.

The foregoing observations, caveats and uncertainties illustrate why researchers and, to a lesser extent, fatigue designers continue to try to unravel the complexities and issues pertaining to metal fatigue, both fundamentally and in engineering practice. However, as the title of this book indicates, it is concerned with Fatigue Management, and in practice this means that FCG is primarily—almost exclusively—of interest.

2.1 Important Aspects of FCG: An Illustrated Example

The following example highlights some important aspects of FCG under realistic VA loading. Figures 2.1, 2.2 and 2.3 present the total fatigue lives followed by Quantitative Fractography (QF)-derived FCG curves for aluminium alloy (AA) 7050-T7451 flat hourglass (waisted) specimens, $Kt = 1.055$, having a finely ground finish, i.e. unpolished. The specimens were tested with an F/A-18 Hornet spectrum at four peak stress levels. Up to five specimens were tested for each stress level, and only the lead crack growth from each is plotted in Figs. 2.2 and 2.3 [5]: each point represents a QF crack growth increment. (Note: these data were further restricted to corner cracks in [6].) Also plotted in Figs. 2.2 and 2.3 are conventional (AFGROW) FCG predictions using standard long-crack growth rate data from handbooks.

The stress–life (S–N) results in Fig. 2.1 provide no insight into why scatter in the total lives occurs, although increasing scatter at lower stress levels is well known. Also, FCG data plotted in a linear–linear fashion in Fig. 2.2 do not reveal any insightful trends. However, plotting the data in a log–linear exponential form in Fig. 2.3 does show clearly identifiable trends, including

  i. The data were approximately exponential, i.e. they followed an approximately straight line over the entire FCG range.
 ii. The slope/gradient (or rate) was proportional to the applied stress, increasing with increasing stress levels.

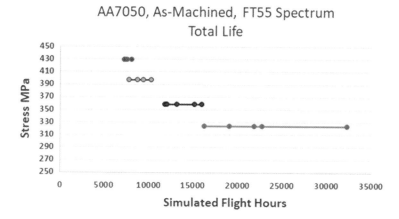

Fig. 2.1 Fatigue test specimen (S–N) data: stress versus Simulated Flight Hours (SFH) (*created by author*)

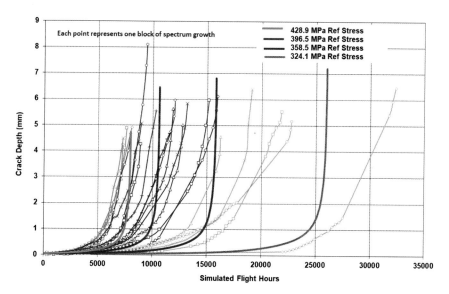

Fig. 2.2 QF-derived FCG for AA7050 specimens loading by an F/A-18 spectrum (FT55) at four peak stress levels. Conventional FCG predictions (AFGROW) using standard long-crack data are also shown, as heavier solid lines (*created by author*)

iii. The variability in the nucleating discontinuity size (i.e. derived from the y-intercepts at zero Simulated Flight Hours (SFH)) largely determined the scatter in the lives for each stress level.

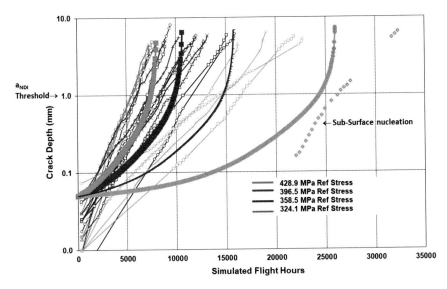

Fig. 2.3 Data from Fig. 2.2 re-plotted exponentially. The AFGROW predictions are shown as heavier solid lines (*created by author*)

iv. In many cases, FCG was measured by QF down to the first spectrum block, i.e. this means little or no nucleation periods or no indications of a threshold for crack growth.

v. There was some variability in the slopes within each stress level set.

vi. The critical crack size at failure was approximately or about the same, probably because the specimens were produced from the same location and orientation within a single billet of material.

vii. One specimen failed from a sub-surface discontinuity, i.e. early cracking in vacuo that resulted in a much longer life for that specimen: see the indication in Fig. 2.3. It was also noted that when the crack broke through to the surface its FGG slope was then similar to those of the other specimens tested at the same stress level.

viii. Up to two-thirds of the total life was spent in growing cracks to a non-destructive inspection (NDI) detectable size (about 1 mm) and mostly within the physically short crack regime.

For the AFGROW predictions it was noted that:

I. The shapes of the curves in Fig. 2.3 were not exponential. This is particularly significant if such predictions were to be used to set inspection intervals, since the amount of predicted FCG in terms of flight hours beyond the 1 mm NDI detectability threshold was typically predicted to be small, whereas in reality the QF results showed that the detectable crack growth could be very significant.

II. Not all the predictions were conservative. This is an often-observed disadvantage when making so-called 'blind' predictions, i.e. not calibrated by representative FCG data for the structures, components or specimens of interest.

III. A relatively large (less realistic) initial flaw depth of about 0.6 mm was required to generate FCG growth.

If for no other reason than clarity, one may conclude that FCG data should be plotted exponentially, i.e. log–linearly. Also, the inherent limitations of total life S–N plots like Fig. 2.1, albeit that the traditional safe-life design basis remains valid, render this approach inappropriate for further consideration in this book.

Statement (iv) about the QF data that there was 'almost no crack nucleation period' is important and deserves some additional comments. In the general literature, there is the view (e.g. [7, 8]) that a significant period of initiation/nucleation is required before FCG occurs, and that this period represents a process of cumulative internal-to-external damage before actual crack formation. Experiments have shown that this process does occur in some metals (usually shown for high-purity aluminium alloys) but only for very carefully prepared and polished specimens. This type of damage is not relevant to aircraft (airframe) structures and components, whose surfaces generally have fine machined/ground topographies, but with occasional minor finishing 'defects/discontinuities', and also some inherent discontinuities, for example, second-phase constituent particles and inclusions in commercial aerospace aluminium alloys and steels.

On the other hand, for certain categories of aircraft components there may be significant fatigue 'initiation' periods. These components are typically carefully finished engine parts or stressed at low levels, e.g. many parts in helicopters, control systems and secondary airframe structures. In these cases, fatigue cracking may eventually begin owing to surface degradation by corrosion, erosion, wear (fretting) and maintenance-induced damage. A special but important case is Foreign Object Damage (FOD) in aeroengine blades and vanes, usually in the fan and compressor sections.

2.2 Analytical Descriptions of FCG: A Brief Summary

In the present context, it is instructive to revisit the well-known LEFM-based Paris equation [9], namely:

$$\frac{\mathrm{d}a}{\mathrm{d}N} = C(\Delta K)^m \qquad (2.1)$$

where a is the crack length at cycle N, ΔK is the stress intensity range (or similitude parameter) and C and m are nominally material constants (note: the units of m and C are interdependent).

Taking the natural logarithms:

$$\ln\left(\frac{da}{dN}\right) = \ln C + m \ln \Delta K \tag{2.2}$$

and integrating for a closed-form solution with a constant width correction factor β, the following are found for the life of a crack:

$$a_f = a_0 e^{c\pi(\Delta\sigma\beta)^2 N_f} \text{ for } m = 2 \tag{2.3}$$

$$a_f = \left[a_0^{\left(1-\frac{m}{2}\right)} + N_f C\left(1 - \frac{m}{2}\right)\left(\Delta\sigma\beta\sqrt{\pi}\right)^m\right]^{\left(\frac{1}{1-\frac{m}{2}}\right)} \text{ for } m \neq 2 \tag{2.4}$$

where a_f is the final crack size and σ is the far field stress.

In this book the long-neglected Eq. 2.3, representing exponential FCG, is of most relevance. This corresponds to the first crack growth equation, attributed to Head [10], an early researcher from the Australian Defence Science and Technology Group, then called the Aeronautical Research Laboratories (ARL). Subsequently, Frost, Dugdale et al. [11, 12] used Head's observation of self-similar (repetitive similarity) crack growth and expanded Head's equation, thereby reporting that crack growth under constant amplitude loading could be described via a simple log–linear relationship (log crack growth vs linear life). Independently, Shanley also proposed the exponential model [13–15], and later the USAF [16, 17] suggested this for small-to-medium length cracks as an approximation of crack growth data available from spectrum fatigue tests.

References

1. J. Glynn, On the causes of fractures of the axles of railway carriages. Minutes of the Proceedings of the Institute of Civil Engineers **3**, 202–203 (1844)
2. The failure of metals by fatigue, in *Proceedings of a Symposium held in the University of Melbourne*, Australia, 2–6 December (1946)
3. C.F. Tiffany, J.P. Gallagher, C.A. Babish IV, Threats to aircraft structural safety, including a compendium of selected structural accidents/incidents. ASC-TR-2010–5002, AFBWP, OH, USA, March (2010)
4. R.J.H. Wanhill, L. Molent, S.A. Barter, Milestone case histories in aircraft structural integrity, in *Comprehensive Structural Integrity*, Second edn. Elsevier Inc., (2023)
5. R.A. Pell, P.J. Mazeika, L. Molent, The comparison of complex load sequences tested at several stress levels by fractographic examination. Eng. Fail. Anal. **12**(4), 586–603 (2005)
6. J.P. Gallagher, L. Molent, The equivalence of EPS and EIFS based on the same crack growth life data. Int. J. Fatigue **80**, 162–170 (2015)
7. S. Suresh, *Fatigue of Materials*, Cambridge University Press, UK, (1998)
8. R.J.H. Wanhill, S. Barter, L. Molent, *Fatigue Crack Growth and Lifing Analyses for Metallic Aircraft Structures and Components.* Springer Briefs in Applied Science and Technology The Netherlands, (2019)
9. P. Paris, F. Erdogan, A critical analysis of crack propagation laws. ASME J. Basic. Eng **85**(4), 528–533 (1963)

10. A.K. Head, The growth of fatigue cracks. Philos. Mag. **44**(7), 925–938 (1953)
11. N.E. Frost, K.J. Marsh, L.P. Pook, *Metal Fatigue* (Clarendon Press, Oxford, 1974)
12. N.E. Frost, D.S. Dugdale, The propagation of fatigue cracks in test specimens. J. Mech. Phys. Solids **6**(2), 92–110 (1958)
13. F.R. Shanley, A theory of fatigue based on unbonding during reversed slip. Rand Corporation Report No. P350. Santa Monica, California, USA (1952)
14. F.R. Shanley, A theory of fatigue based on unbonding during reversed slip - revised. Rand Corporation Report No. P350. Santa Monica, California, USA (1953)
15. F.R. Shanley, A proposed mechanism of fatigue failure, in *Proceedings of the Colloquium on Fatigue/Colloque de Fatigue/Kolloquium über Ermüdungsfestigkeit*. Stockholm, 25–27 May 1955 Springer, Berlin and Heidelberg, (1956), pp 251–259
16. S.D. Manning, J.N. Yang, USAF durability design handbook: guidelines for the analysis and design of durable aircraft structures, AFWAL-TR-83-3027, Wright-Patterson Air Force Base, Ohio, USA (1984)
17. A.P. Berens, P.W. Hovey, D.A. Skinn, Risk analysis for aging aircraft fleets. Analysis, vol 1, WL-TR-91-3066, Dayton University Ohio Research Institute, Ohio, USA, (1991)

Chapter 3
Some Design Considerations

3.1 Existing Airframe Design Standards

Aircraft design standards provide the requirements and/or guidance to assist a manufacturer in providing a safe and durable airframe for a specific period of operation or an inspection-free period. The standards also advise on the design and in-service tasks or general aspects/elements that are required to achieve and maintain structural airworthiness. The elements of in-service aircraft structural integrity programmes (e.g. see Table 1 in reference [1]) are summarized in Fig. 3.1. This shows both the interdependence of the elements and the need for constant monitoring and review. The standards advise on the choices of fatigue-resistant materials and the use of a so-called building-block approach to establish the design lifetime. The approach consists of

(a) Choice of materials.
(b) Establishing the basic fatigue properties of the materials using specimens and coupons tested under constant amplitude (CA) loading.
(c) Testing components that incorporate some of the detailed design features, including (i) joints and geometrical discontinuities such as notches and (ii) surface finishes (coatings) that may affect the fatigue performance, and tests under variable amplitude (VA) loading derived from the assumed design spectrum.
(d) An FSFT of a representative airframe using distributed loads derived from some of the points-in-the-sky of the design spectrum. Normally these points are chosen to ensure that the structure is subjected to the flight conditions considered to be the most fatigue-damaging. If the test article achieves the design testing targets, preferably including a residual strength demonstration, the FSFT then determines or validates the useful (unfactored) life of the airframe.
(e) Choice of an appropriate safety factor to cover the potential variability in fatigue behaviour of the materials, components and structures (i.e. a scatter factor, which may be based on so-called engineering judgement), and in some instances an

© The Author(s), under exclusive license to Springer Nature Singapore Pte Ltd. 2024
L. Molent, *Aircraft Fatigue Management*, SpringerBriefs in Applied Sciences
and Technology, https://doi.org/10.1007/978-981-99-7468-9_3

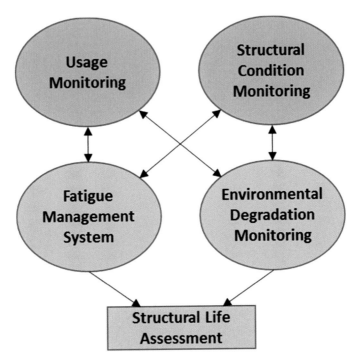

Fig. 3.1 Schematic showing the elements for in-service management of aircraft structures: courtesy of Group Captain R. Singh, Royal Australian Air Force (RAAF)

additional factor owing to the anticipation that the fleet operational loads will be more severe than those derived from the design spectrum. This possibility/probability will exist because of past experience with similar aircraft types.

The design standards require the airframe to be tested by the manufacturer to a multiple of the anticipated design life. Since an FSFT is expensive, the manufacturer is generally inclined to conduct the test no longer than the minimum required duration (see Sect. 3.6). In addition, the high costs of an FSFT make it necessary to apply a load history derived from a sub-set of the assumed and/or anticipated in-service load conditions, and also for testing to be done for a limited sub-set of aircraft weights and configurations.

Owing to the foregoing restrictions and limitations, the design standard requirements sometimes prove to be inadequate for enabling the airworthiness manager to establish the airframe life for the actual configurations, roles and environment in which the fleet operates. Specifically, there are a number of factors that can give rise to this inadequacy. These factors are

A. The operational spectrum loads can be (and often are) significantly different from the design spectrum for the following reasons:

    a. The aircraft are operated under new conditions. For example, new generation aircraft are frequently operated more severely than the legacy aircraft they replace, because of improved performance and more demanding (upgraded) missions.

    b. An aircraft's weight generally increases (or varies) with time in service, owing to factors such as mission system upgrades, new weapons, repairs, painting and re-painting, and the accumulation of unavoidable detritus (e.g. entrapped dirt and water, build-up of greases, paint, etc.).

B. Fatigue cracks are normally detected during the course of an FSFT, and these are usually repaired so that the overall test airframe meets its design objectives. However, in some cases, the repairs are not subjected to sufficient testing to adequately address their effectiveness and durability. This forces the fleet managers to adopt conservative approaches to their implementation in the fleet. Also, repairs may have parasitic effects on nearby structure, e.g. local stiffening that changes the load paths. Such effects have the potential to generate new and unanticipated fatigue hot spots.

C. The FSFT may demonstrate adequate lives for those areas that had been tested with representative applied loads, but it is often not possible to conduct a test that is representative in all areas. Thus when cracks form in non-representatively loaded areas, it may be necessary to establish the correct FCG behaviour and residual strength level for these cracks by analysis, possibly verified by additional coupon testing.

D. Modern aircraft often remain in service longer than initially anticipated or required, either because of budgetary considerations or because a replacement aircraft is not yet available. This is why many aircraft undergo Service Life Extension Programmes (SLEPs). A major problem with SLEPs is that because the original FSFT would have been of limited duration, it is usually impossible to predict the next 'weak-link' in the structure during the extended life. For this reason, follow-on FSFTs are normally warranted, despite the considerable costs involved.

3.2 Conventional Fatigue Analyses

Many of the problems and scenarios summarized in the previous chapters of this book require fatigue analyses to estimate the service lives of aircraft. Conventional fatigue analyses are empirical and based on variants of the stress–life, strain–life or FCG models. Despite many decades of research and development, significant challenges remain in using these models to obtain accurate and consistently conservative predictions of the lives of metallic structures subjected to VA loading, see, for example, [2]. Two examples to show the importance of these challenges are briefly mentioned here.

Firstly, it is well known that the fatigue lives of metallic structures are very sensitive to changes in the applied stress levels. As an illustration, the change in fatigue lives of unnotched components and coupons is approximately proportional to the third power (cube) of the change in stress amplitude, (see, for example, [3] and Sect. 5. 1). This strong dependence, coupled with the uncertain reliability of the available fatigue prediction tools, can lead to highly conservative service life estimates when analytically interpreting the original FSFT results. A specific illustration concerns the F/A-18 Hornet aircraft when in service with the RAAF. Soon after entering service the operational spectra were found to be significantly different from the design spectrum. This finding, and a lack of confidence in the ability to translate the design FSFT results to the actual operational usage, resulted in a follow-on FSFT under representative RAAF usage [4, 5]. This follow-on test then became the basis for the through-life fatigue management of the aircraft.

Secondly, conventional FCG analyses are unable to describe the behaviour of physically small cracks (less than about 1 mm in depth). These cracks are very important for high-performance aircraft, since they typically account for one-third to two-thirds of the airframe's life [6, 7]. Hence, while conventional FCG models using initial surrogate flaw sizes may by adequate for design, they are generally inappropriate for service aircraft or FSFT failure analyses. More or less the same conclusion was drawn in the discussions of Figs. 2.1 and 2.2.

The innovative non-LEFM fatigue analysis tools used to maximize the value of follow-on FSFTs and cope with the small crack problem are described in Chap. 5, as was mentioned in Chap. 1.

3.3   Constant Versus Variable Amplitude Loading

CA and VA loading can produce markedly different FCG fracture surface topographies [7, 8]. Two examples will be given here. Firstly, Fig. 3.2 shows a change in crack path direction when the loading is varied between VA and CA, and back again. Secondly, Fig. 3.3 shows changes in fracture topography for a simple CA + underload sequence and in-between CA loading.

In fact, it is a general result that VA loading, particularly containing underloads, e.g. the unavoidable and important ground-air-ground (GAG) cycles for aircraft, results in flatter fatigue fracture surfaces. While this is actually serendipitous for QF, it means that the FCG resistance decreases [7]. Such an effect is not accounted for when using conventional FCG models, which use standard CA long-crack growth data. For this reason, the novel fatigue lifing methods described in this book rely predominantly upon actual VA FCG data rather than calculated VA FCG curves derived from conventional CA-based lifing tools.

An additional significant aspect is that these topography changes enable selecting representative VA load histories and sequences that provide fracture surface markers for QF [8–11]. QF is an essential tool for determining the FCG behaviour of short cracks, as illustrated by Fig. 2.3. QF is also very useful for the post hoc determination of crack growth in FSFT and service components, whether or not they have been subjected to intermittent inspections to detect and follow cracking.

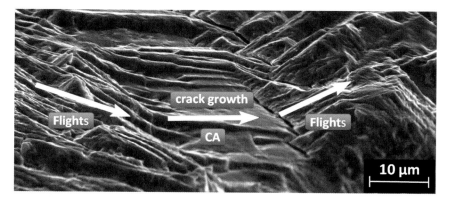

Fig. 3.2 SEM image of AA7050-T7451 fracture surface changes when the applied loading switches from repeated single flight loading to a band of constant amplitude (CA) loading and back to single flight loading [8] (created by author and colleagues)

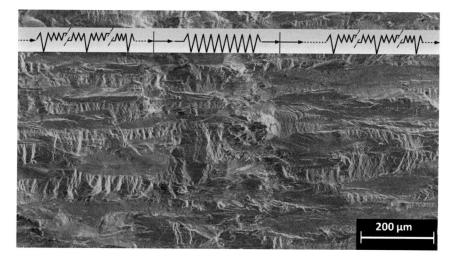

Fig. 3.3 SEM anaglyph (stereopair) image of AA7075-T7351 fracture surface changes when the applied loading switches from repeated simple VA loading to a band of constant amplitude (CA) loading and back to simple VA loading: courtesy of R. J. H. Wanhill. This image is best viewed using special blue-and-red glasses

3.4 Relative Spectrum Severity

A common requirement of airframe fatigue analysis is to determine the relative fatigue severity of two VA load spectra when they are applied to the same components or structures at a given stress level. An obvious example is when the load history used in an FSFT is later found to be unrepresentative of the (average) fleet usage. It is then necessary to (a) determine the extent to which the actual usage is more, or less severe, than the test load history and (b) reassess the FSFT results so that they may be used to represent the service usage. Although conventional fatigue prediction tools/models are available for this task, they usually contain uncertainties that are difficult to remove without extensive and additional coupon testing dedicated to this task.

Even coupon testing for this purpose is subject to variability and thus uncertainty. Coupons may be manufactured from different material batches or areas of the same material billet, leading to (some) additional scatter in results. In a novel approach, a single-coupon test programme was developed to assess the relative FCG severity of two load histories derived from two spectra. The load histories were applied to the coupon sequentially, and with numerous repeats of the sequencing, such that they covered FCG from small naturally nucleating cracks to failure. The FCG data were subsequently obtained using QF. This enabled direct comparisons of the FCGRs from the two spectra at different stages of the crack growth life, without concerns for previously unacknowledged or unrecognized sources of scatter [12].

It is critically important that each spectrum is applied at the correct stress level. Each spectrum is likely to have different maximum peak loads, and it would be incorrect to test each spectrum at one maximum stress level. For this reason, a reference stress (σ_{REF}) is chosen for a common load level (e.g. 6g's) and the spectrum stresses are scaled against this reference stress.

3.5 Individual Aircraft Fatigue Tracking

The most basic and minimum objective of fatigue and structural integrity management is to ensure that the life of type of an aircraft safely meets the operator's planned withdrawal date. In the first instance, this should be achievable under normal operating loads and within approved flight limitations (and without unacceptable deformation or failure of the airframe) corresponding to a desired level of safety or probability of failure (POF): in short, the ensuring of structural integrity. The adopted philosophy and methodology for achieving this basic objective depends on a number of factors, such as the ability to inspect and repair or replace particular components, and the consequences of complete failure of a component.

Knowledge of the applied loads is paramount to the basic objective, as it is well known that fatigue lives at specific structural locations are sensitive to the actual load history and, as mentioned in Sect. 3.1, the overall stress levels. Therefore, good management of fatigue or structural integrity requires establishing the service load history and environmental conditions by monitoring critical/key locations in the airframe. Given the typical variability of loading between individual aircraft [13–15], it is recommended that each aircraft in a fleet be monitored. In fact, it is now mandatory for all combat aircraft to be fitted with airborne fatigue monitoring systems (sometimes referred to as Structural Health and Usage Monitoring or Prognosis Health systems (HUMS, SHM, PHM, etc.), see, for example, Ref. [16]. These systems typically collect operational data that are used for calculating the overall remaining safe-life of the airframe and/or the inspection intervals to ensure continued structural integrity.

The techniques used in individual aircraft fatigue monitoring programmes and the current systems and practices have been comprehensively examined and reviewed in [14]. It was found that there is little uniformity in the fatigue management practices of operators, and that many aspects of the fatigue management process have been overlooked by some structural integrity managers. Also, very few of reviewed papers in [14] described the philosophy or objectives in choosing specific monitoring systems. Based on experience from Australian fatigue monitoring programmes, this review discussed the issues of strain gauge utilization and calibration, the collection of flight parameter data, data integrity and comparisons with fatigue test results and predictions from fatigue damage models. The discussion enabled pointing out some of the potential (and actual) pitfalls in the techniques, systems and practices, and this led to recommended objectives and requirements for future fatigue monitoring programmes. The recommended objectives and requirements are summarized here according to the review's numbering system (see [14]):

2.2 The aim of the fatigue monitoring system is to ensure that an aircraft does not exceed the damage limits demonstrated in the FSFT or design and with suitable safety factors applied. Ideally, the system should provide the fleet manager with a timely non-dimensional parameter, e.g. Fatigue Life Expended Index (FLEI). An FLEI = 1.0 is a certification limit, see [17], that indicates the consumed amount of the safety-factored FSFT-demonstrated fatigue life. A novel approach using an FCG model based on physically short crack data is described in [18] and used for the RAAF F/A-18 Hornet.

2.3 A primary means of assurance is translation of the individual aircraft load histories into fatigue damage at critical/key locations that are compared to the FSFT-demonstrated FLEI. The USAF uses a different method, the Equivalent Flight Hours (EFH) approach [19]. Here the usage severity for each aircraft in the USAF fleet is compared to a reference usage (normally the design spectrum for the location of interest) to assess the structural health and maintenance requirements for each 'tail number' in the fleet.

2.4 Models used in design or to set inspection intervals must be validated or calibrated against FSFT results. The model(s) used to translate the service load histories into fatigue damage must be shown to be consistent and suitably conservative in its/their coverage of the range of actual and anticipated fleet usage [17, 20].

2.5 The primary load measuring method should be strain gauges (supplemented with accelerometers for aerodynamic-buffet-dominated locations). These gauges must be calibrated, see, for example, [20], to a known load level and must be located at the same locations as those used during the FSFT.

2.6 The gauges must be placed in areas of uniform stress and not at potential fatigue hotspots. These hotspots are likely to experience significant stress gradients, e.g. owing to the presence of notches or structural discontinuities, such that their use in load monitoring is questionable. It should be noted that placing strain gauges at predicted hotspots on FSFT articles is also not advisable for the following reason. Strain gauging requires good surface preparation, which in many cases requires careful polishing to ensure gauge durability. The polishing is likely to remove potential surface-related discontinuities (see Chap. 6) such that fatigue cracking may be delayed or even not occur during service.

2.7 The means of establishing data and system integrities must be demonstrated and verified. A post-mid-life fleet aircraft teardown and inspection is a useful means of assessing the veracity of the entire aircraft structural integrity plan.

2.8 The usage results must be communicated to fleet managers in a meaningful manner.

3.6 Full-Scale Fatigue Test Considerations

Design requirements generally mandate that aircraft manufacturers conduct an FSFT to a specified multiple of the required design life. Modern aircraft are designed to one or more of the following criteria:

- SAFE-LIFE: the structure has been evaluated to assure that it will withstand the repeated loads of variable magnitude expected during its service life without detectable cracks.
- FAIL-SAFE: the structure has been evaluated to assure that catastrophic failure is not probable after failure or obvious partial failure of a single, principle structural element.
- DAMAGE TOLERANT: the structure has been evaluated to ensure that should serious fatigue, corrosion, or accidental damage occur within the operational life of the aircraft, the remaining structure can withstand reasonable loads without failure or excessive deformation until the damage is detected. Generally, FSFT spectrum is considered to represent mean fleet usage.
- DURABILITY: The structure shall be adequate to resist fatigue cracking, corrosion, thermal degradation, delamination, and wear during operation and maintenance. Generally, FSFT spectrum is considered to represent severe fleet usage.

The FSFTs are used to validate the design and the design tools including FCG models.

As mentioned in Sect. 3.1, these requirements may not be sufficient to ensure the long-term structural integrity of a fleet of aircraft.

Rather than testing to a specified multiple of the required life, it is generally recommended that a representative FSFT be continued for as long as economically feasible. Its primary purpose then becomes the identification of all potentially fatigue-critical locations, thus revealing all 'weak-links'. Furthermore, an extended FSFT could provide the necessary information and data to conduct a SLEP (if required, which is very likely to be the case, as pointed out earlier).

Another benefit of conducting an FSFT in this manner is the determination of an airframe's potential to suffer Multiple-Site fatigue Damage (MSD) and/or

Widespread Fatigue Damage (WFD), and also Multiple-Element fatigue Damage (MED). This is now a design requirement [21, 22]. The critical aspect of these fatigue-cracking scenarios is that it could be possible that cracks interact to compromise the residual strength of the structure far more than if they were effectively isolated (see Sect. 3.8). These multiple-crack conditions and scenarios are truly life-limiting, since the actual or potentially large numbers of cracks can overwhelm the guarantee of safety by inspection. These conditions also represent the limit of FSFT validity [21].

At the completion of fatigue testing, a residual strength test should be done. This has the purpose of determining the significance of any possible multi-crack conditions. After the residual strength test, the test article should undergo a thorough teardown, including inspections to detect all instances of cracking, subject to the detectability thresholds of the inspection techniques used. N.B.: It is important to budget for the cost of teardown and subsequent QF of significant cracks. This is labour intensive, requiring dedicated equipment and much expertise.

It is good practice to establish the condition of the fleet through a post-mid-life teardown inspection of a fleet aircraft. Sometimes an airframe may become available for this purpose due to unserviceability or damage. As small cracks below NDI thresholds are likely to exist, then a way of enhancing or growing these to a detectable size is most advantageous. Enhanced fatigue testing and teardown are a method of rapid, destructive structural experimentation devised to unlock key data and evidence required for fleet Structural Integrity management. An enhanced teardown (e.g. [23]) differs from a traditional FSFT in two ways:

- The test article must be a retired fleet asset with the structural degradation arising from production/in-service usage 'locked in' for identification through testing and forensics; and
- Limitations in loads fidelity are accepted from the onset if testing remains fit for purpose and can be simplified/accelerated as a result. The condition of the fleet then supports (or otherwise) the Aircraft Structural Integrity Plan (ASIP) e.g. [24].

3.7 Residual Strength Criterion

Because cracks will always exist in an airframe, albeit generally below the threshold of detectability of current NDI techniques, see Chap. 4, the structure must be capable of safely resisting factored operational loads as it approaches the end of its life. The factor used to scale the Design Limit Load (DLL) should be less than 1.5, which is used in the design phase to define the design ultimate load = 1.5 X DLL. The reason for using a reduced factor, less than 1.5, is that individual aircraft monitoring systems would (or should) have established the peak operational loads during service. Notwithstanding the availability of this information, a specified reduction factor is still applied to the Operational Limit Load (OLL) to account for an unanticipated extreme loading.

Traditionally, the RAAF methodology for lifing aircraft primary structures required establishing the fatigue test life, under representative loading, of a full-scale structure or major component to a residual strength (RS) requirement of 1.2 X DLL or the OLL (whichever was the greater) without failure, debilitating deformation or stiffness loss resulting in potential flutter issues. Whether the test structure fails below 1.2 X DLL or survives, a method is required to determine the equivalent fatigue life defined by a location's ability to achieve and survive 1.2 X DLL with cracking present. In effect, this means that the test time (equivalent RS life) to the critical crack length or depth (a_{RST}) at the 1.2 X DLL point is required. There are three basic cases to consider:

a. If a particular crack fails the structure below 1.2 X DLL, its fatigue life is reduced to a time at which it would have reached a calculated RS test (RST) critical crack size, a_{RST}.
b. For those cracks that survive the RST load some assessment of the amount of remaining FCG life may also need to be done.
c. During the fatigue cycling part of an FSFT, it is often necessary to ensure the test article's survival by removing or modifying cracked locations when the cracks are smaller than the size at which they would have failed under an RST. These locations subsequently become the subject of fleet action prior to the desired life demonstrated by the overall FSFT. The lives at these locations can be reassessed in the following way. If the pre-modified FCG of the locations can be established, then the FCG plots can be extrapolated to the relevant a_{RST} values to establish what may be considered as virtual test points. These virtual test points are then used for lifing purposes.

More details of a method of dealing with all these cases are summarized in Chap. 7.

3.8 Multiple-crack Fatigue Damage

As mentioned in Sect. 3.6, multiple-crack fatigue damages where cracks have the potential to interact are variously defined as MSD, WFD and MED. These multi-crack conditions are the major ones affecting the structural longevity of an aircraft or fleet of aircraft, and have given rise to the term 'ageing aircraft' or less flatteringly, 'geriatric aircraft'.

Recognition of multiple-crack fatigue damage and the ageing aircraft problem is commonly attributed to the Aloha Airlines accident [25], though this accident does not necessarily represent the first occurrence of this type of damage. The active cause of this accident was MSD. However there were other, more basic causes [25] including the poor choice of structural adhesive, exceeding safe-lives, poor maintenance and fastener hole designed-in 'knife-edges'. MSD occurred when fatigue cracks nucleated independently at many adjacent fastener holes in fuselage panel lap joints. Eventually the cracks were able to interact under normal loading conditions such that they suddenly linked to form a large crack that defeated the fail-safety of the fuselage. N.B.: For fail-safety, the structures are designed to ensure that catastrophic failure is improbable after the failure or obvious partial failure of a single primary structural element in a multi-element structure. This damage should be detected during normal operational maintenance, i.e. it should be obvious.

To adequately manage the problems of MSD, WFD and MED, it is necessary to understand the phenomena and have predictive tools that provide the engineering framework and methodology for maintaining continued airworthiness. In this context, a significant contribution has been made to the problem of MSD. It was first shown [26] (and subsequently checked by re-analysis [27, 28]) that adjacent cracks, typical of MSD in simulated fuselage lap joints, grow and act independently until the point at which the ligaments between neighbouring crack tips undergo complete yielding. Subsequently, the ligaments fail by overload, leading to link-up of the cracks.

This behaviour was demonstrated using a relatively simple lap-joint specimen configuration that adequately reproduced the MSD FCG rates seen in service aircraft. (These results were available prior to the Aloha incident and were subsequently provided to Boeing to support continuing airworthiness of the fleet.) In these tests, a worst-case scenario was assumed, namely, a non-bonded, upper plate countersunk configuration with small crack-starter notches that in most cases were removed when reaming the holes to 4.039 mm diameter. Each specimen consisted of two 1.02-mm-thick aluminium alloy 2024-T3 clad sheets fastened with three rows of 3.97 mm in diameter BACR15CE-5, 100° shear head countersunk rivets. The spacing between the rivets was approximately 25.4 mm. Also, the specimen widths were chosen to coincide with the typical distance between tear straps of a Boeing commercial aircraft. The specimens were tested in tension to give a peak remote plate stress of 92 MPa, with $R = 0.05$, in an electrohydraulic test machine. Figure 3.4 shows a series of images taken from a test (two rivets on the top row) to demonstrate the cracks linking up via monotonic plastic deformation at the crack tips.

Understanding this link-up behaviour makes it possible to analyse MSD and WFD. However, it is still necessary to consider the potential for a large number of cracks to overwhelm the normal scheduled inspections intended to ensure safety.

A further significant and fundamental finding from this MSD study was the experimentally measured stress concentration effect of a hole containing a neat fit fastener (e.g. a rivet). The stress concentration effect was approximately half the effect given in the classic textbook open-hole solution [26]. In other words, the stress concentration factor, Kt, for the neat-filled hole was only about 1.5 rather than the conservative open-hole value of 3 that is conventionally used.

3.9 Scatter in Fatigue Performance

We have improved the understanding of fatigue behaviour of metal airframes to the point where for a fixed loading spectrum, stress level, environment and material, it is now known that the scatter in the fatigue performance of monolithic metallic

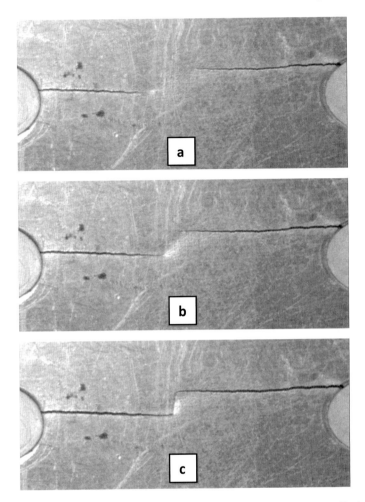

Fig. 3.4 Sequence of images from a lap-joint specimen cycled under constant amplitude loading [28]. **A** is prior to monotonic plastic deformation at the crack tips. **B** shows signs of plastic deformation before the crack tips link-up. **C** shows full plastic collapse (*Source* author)

airframe components is governed by the variability in the metal's material properties and manufacturing quality. (Note that variations in aircraft usage will also lead to significant scatter in fatigue lives, but this is addressed separately through fatigue load monitoring.) These material and manufacturing variabilities can be quantified, see [29], by gaining an understanding of the scatter in the parameters shown in Table 3.1. Parameters 1–3 in Table 3.1 are considered to define the aircraft's build quality from a fatigue perspective. Parameters 4–7 define the metal's property variability. Thus if the variation in these latter parameters can be adequately quantified, then a more accurate estimate of the scatter in an alloy's fatigue performance can be made.

Table 3.1 Sources of manufacturing and material (alloy) fatigue variability

Parameter	Description
1	Population density/size distribution of nucleating discontinuities that lead to fatigue cracking
2	Stress concentrations leading to inter-aircraft variations in local stress
3	Fit-up (assembly) stresses or residual stresses
4	Material cyclic stress intensity threshold (if it really exists)
5	Fracture toughness of the material
6	Crack nucleation and/or initiation period (including sub-surface nucleation effects)
7	FCG rates of cracks in the material when being measured at nominally similar stress locations

A detailed investigation into fatigue scatter considered the AA 7050-T7451 thick plate used in the F/A-18A/B centre fuselage [29]. The fatigue scatter was quantified by estimating the standard deviation of log(10) fatigue lives from FSFT and coupon fatigue tests. The focus of the investigation was on low stress concentration (Kt) areas of the centre fuselage where the manufacturer's etched surface finish was present. Most of the FSFT and coupon tests were applicable to these conditions, i.e. it was expected that there would be low fatigue variabilities from at least some of the parameters listed in Table 3.1.

There was a total of 26 FSFT and 177 coupon fatigue lives, whereby the coupons were grouped in such a way so as to replicate the conditions expected at fatigue-critical locations in the centre fuselage. The standard deviations of the log(10) fatigue lives calculated for the individual groups of bulkhead failure locations and coupons ranged between 0.065 and 0.12. Also, there was no discernible trend of increasing fatigue scatter with increasing fatigue life. This result contrasts to 'received engineering wisdom' and shows the importance of minimizing the fatigue variabilities of the manufacturing and material parameters.

References

1. MIL-STD-1530D (W/ Change-1), Department of Defense Standard Practice: Aircraft Structural Integrity Program (ASIP), October (2016)
2. J. Schijve, Fatigue of structures and materials in the 20th century and the state of the art. Int. J. Fatigue **25**, 679–702 (2003)
3. S. Barter, L. Molent, N. Goldsmith, R. Jones, An experimental evaluation of fatigue crack growth. Eng. Fail. Anal. **12**(1), 99–128 (2005)
4. D.L. Simpson, N. Landry, J. Roussel, L. Molent, A.D. Graham, N. Schmidt, The Canadian and Australian F/A-18 international follow-on structural test project, in *Proceedings ICAS 2002 Congress*, Toronto, Canada, September (2002)
5. L. Molent, S. Barter, P. White, B. Dixon, Damage tolerance demonstration testing for the Australian F/A-18. Int. J. Fatigue **31**, 1031–1038 (2009)
6. L. Molent, S.A. Barter, A comparison of crack growth behaviour in several full-scale airframe fatigue tests. Int. J. Fatigue **29**, 1090–1099 (2007)

7. R.J.H. Wanhill, Damage tolerance engineering property evaluations of aerospace aluminium alloys with emphasis on fatigue crack growth, NLR TP 94177, The Netherlands (1994)
8. P. White, S. Barter, L. Molent, Observations of crack path changes caused by periodic underloads in AA7050-T7451. Int. J. Fatigue **30**, 1267–1278 (2008)
9. R.J.H. Wanhill, S. Barter, L. Molent, *Fatigue Crack Growth and Lifing Analyses for Metallic Aircraft Structures and Components*. Springer Briefs in Applied Science and Technology, The Netherlands, (2019)
10. S.A. Barter, L. Molent, R.J.H. Wanhill, Marker loads for quantitative fractography of fatigue cracks in aerospace alloys, in *Proceedings ICAF 2009*, Rotterdam, 27–29 May (2009)
11. R.J.H. Wanhill, N. Goldsmith, L. Molent, Quantitative fractography of fatigue and an illustrative case study. Eng. Fail. Anal. **19**, 426–435 (2019)
12. S. Barter, B. Dixon, L. Molent, Assessing relative severity using single fatigue test coupons. Eng. Fail. Anal. **16**(3), 863–873 (2009)
13. Ministry of Defence, Defence Standard 00-970 Issue 1, Design and Airworthiness Requirements for Service Aircraft. Aeroplanes, vol. 1, Amendment 13, December (1994)
14. L. Molent, B. Aktepe, Review of fatigue monitoring of agile military aircraft. Fatigue Fract. Eng. Mater. Struct. Fract. Eng. Mater. Struct. **23**, 767–785 (2000)
15. L. Molent, J. Agius, Structural health monitoring of agile military aircraft, in *Encyclopedia of Structural Health Monitoring*, eds. by C. Boller, F. Chang, Y. Fujino John Wiley & Sons, Ltd., (2008)
16. Aircraft Structures, Joint Service Specification Guide JSSG-2006, US Department of Defense (1998)
17. T. Dickinson, L. Molent, Validation of fatigue damage models used for F/A-18 life assessment using fatigue coupon test results, DSTO-TR-0940, Defence Science and Technology Organisation, Melbourne, Australia, February (2000)
18. P. White, D. Mongru, L. Molent, A crack growth based individual aircraft monitoring method utilizing a damage metric. Struct. Health Monit.. Health Monit. **17**(5), 1178–1191 (2018)
19. Methodology for determination of equivalent flight hours and approaches to communicate usage severity, USAF Structural Bulletin EN-SB-09-001, June (2006)
20. B. Aktepe, K.P. Hewitt, L. Molent, R.W. Ogden, Validation of RAAF F/A-18 analytical fatigue strain sensor calibration techniques using an aircraft ground calibration procedure, DSTO-TR-0641, Defence Science and Technology Organisation, Melbourne, Australia, January (2000)
21. Airworthiness Assurance Working Group for the Aviation Rulemaking Advisory Committee's Transport Aircraft and Engine Issues Group. Recommendations for Regulatory Action to Prevent Widespread Fatigue Damage in the Commercial Airplane Fleet. Final report, Revision A, 29 June (1999)
22. Aircraft Structural Integrity Program (ASIP), MIL-STANDARD-1530D, USA, August (2016)
23. L. Molent, B. Dixon, S.A. Barter, G. Swanton, Outcomes from the fatigue testing of seventeen centre fuselage structures. Fatigue **111C**, 220–232 (2018)
24. L. Molent, V.T. Mau, Verification of an airframe fatigue life monitoring system using ex-service structure, Eng. Fail. Analysis **83**, 207–219 (2017)
25. Aloha Airlines, Flight 243, Boeing 737–200, N73711, Near Maui, Hawaii April 28, 1988, NTSB Report AAR-89-03 (1989)
26. L. Molent, R. Jones, Crack growth and repair of multi-site damage of fuselage lap joints. Eng. Fract. Mech. Fract. Mech. **44**(4), 627–637 (1993)
27. R. Jones, L. Molent, S. Pitt, Understanding crack growth in fuselage lap joints. Theor. Appl. Fract. Mech.. Appl. Fract. Mech. **49**, 38–50 (2008)
28. U.H. Tiong, R. Jones, L. Molent, J. Huynh, Analysis of the growth of short fatigue cracks in fuselage lap joint, in *Structural Integrity and Failure Conference* (Institute of Materials Engineering Australasia Ltd., 2006), pp. 33–39
29. B. Dixon, L. Molent, S. Barter, A study of fatigue variability in aluminium alloy 7050–T7451. Int. J. Fatigue **92**(1), 130–146 (2016)

Chapter 4
Lead Cracks and an Example of Fatigue Lifing

4.1 Behaviour of Lead Cracks

Many years of QF on metallic airframe components from service and FSFTs have consistently shown that the dominant fatigue cracks (those first leading to failure) grow in an approximately exponential manner (e.g. [1–28]). These QF-based observations cover the nucleation of cracks and their early growth from a few micrometres to many millimetres in size. It appears that these 'lead cracks' commence growing shortly (almost immediately) after the aircraft enter service. Furthermore, these cracks usually nucleate from surface/near-surface production-induced discontinuities or, less frequently, from inherent material discontinuities, e.g. Refs. [8–29], see Chap. 6 also. N.B.: It is worth noting with respect to material choices for aircraft structures that the early growth of damage in composite laminates also appears to be exponential [30, 31]. Further, that while all current FCG models are empirical in nature, fractal analyses appear to provide some fundamental basis for exponential FCG [28].

These observations have resulted in an innovative aircraft lifing methodology called the lead crack fatigue lifing framework (LCFLF) for use with metallic aircraft structures. This methodology has been developed and implemented as an additional tool in determining aircraft component fatigue lives for several aircraft types in the RAAF fleet, for example, see Refs. [32–34].

If a particular area of a structure is susceptible to cracking, it is possible that a number of cracks will nucleate and grow. The fastest growing crack in this area is the (local) dominant or lead crack. There will most likely be a number of lead cracks within the entire structure, and one of these will ultimately cause failure of the structure unless appropriate measures (repair or replacement) are taken.

The general characteristics of lead cracks can be summarized as follows:

(1) They commence growing in high stress areas from (usually surface) discontinuities soon after the aircraft is introduced into service. This implies that the cyclic stress intensity threshold (ΔK_{THRS}) in these areas is low or non-existent.

© The Author(s), under exclusive license to Springer Nature Singapore Pte Ltd. 2024
L. Molent, *Aircraft Fatigue Management*, SpringerBriefs in Applied Sciences and Technology, https://doi.org/10.1007/978-981-99-7468-9_4

(2) Irrespective of local geometry, they grow approximately exponentially with time, i.e. log a (the crack depth or length) versus linear life or cycles, if:

  a. Little error is made when assessing the effective crack-like size of the fatigue-nucleating discontinuity. For example, an error in underestimating the effective crack-like size of the nucleating discontinuity will cause an apparent departure from exponential behaviour over a short period of early FCG.

  b. The crack does not grow into an area of significant change in component thickness or geometry, particularly when the crack depth is small in comparison to the component thickness/width, and either before or after a change in thickness.

  c. No significant load shedding occurs, i.e. the crack is not unloaded as the component either loses stiffness and sheds load to surrounding members or grows towards a neutral axis owing to the predominance of loading by bending.

  d. The crack does not encounter a significantly changing stress field, e.g. it does not grow into or from a region containing residual stresses.

  e. The crack is not retarded by very occasional and very high loads (usually more than 1.2 X the peak load in the spectrum).

  f. The small fraction of life involved in fast fracture or tearing near the end of the fatigue life is ignored (in modern alloys significant tearing usually begins during the highest load shortly before failure). In addition, the general failure/residual strength criterion of 1.2 X DLL (Design Limit Load), as required by DEF STAN 970 or the Maximum Spectrum Stress (JSSG-2006), would tend to eliminate this final period of FCG from fatigue lifing calculations.

Within the limits given in (2) above, many observations of lead cracks have led to the following generalizations:

(3) A change of the geometry factor β (which depends on the ratio of the crack length to width and the component geometry) with crack depth/length does not appear to significantly influence the FCG behaviour. For low Kt features, most of the life is spent when the crack is physically small so that β does not change much. However, even when a lead crack starts at an open hole, where β changes rapidly, it still appears to grow in an approximately exponential manner. (This is not to say that there is no geometry influence: under the same net section stresses, cracks from open holes grow faster than cracks from low Kt details.) The unexpectedly small or negligible influence of a high β gradient on the shape of a lead crack curve requires further research.

(4) Typical nucleating discontinuities and defects for combat aircraft metallic materials (e.g. high-strength aluminium alloys) are approximately equivalent to a 0.01-mm-deep fatigue crack (see, for example, [35–37]). Thus, a 0.01mm deep flaw is generally a good starting point for FCG assessment in these materials (To obtain good predictions using conventional FCG models from such a small

initial crack size will need small crack growth rate data that have been validated for the fatigue predictive tool to be used. While such data are now becoming available, for example, see [38, 39], in the first instance the LCFLF does not require it).

(5) Despite limitation 2)(d) above, lead cracks often appear to grow exponentially within residual stress fields, albeit at faster or slower acceleration rates depending on the sign of the residual stress.

(6) If high loads that retard FCG occur periodically and fairly regularly throughout the life, then the average FCG behaviour will still be exponential.

(7) Although the critical crack size should be readily calculable, it has been observed that the typical critical crack depth for highly stressed areas in combat aircraft metallic components is usually about 10mm. While many cracks in highly stressed components may exceed this size at final failure, it is usually observed that significant tearing has occurred well before failure. In other words, the FCG period beyond a crack size of 10mm may be generally neglected. This can be a convenient approximation for use in life assessment. The same appears to hold for military transport aircraft in the absence of load shedding or slant cracking, e.g. [32].

4.2 Exponential FCG Generality and a Simplified LCFLF Lifing Method

Approximately, exponential FCG behaviour from small discontinuities of approximately 0.01 mm in depth appears to be the norm for lead cracks. It is in fact a common behaviour for numerous materials used in different aircraft types [1, 4, 25]. Exponential FCG is also observed for (a) a range of load histories (e.g. tension-dominated, compression-dominated, combat aircraft, transport aircraft); (b) various structural configurations, e.g. [4, 8, 25], including fuselage lap-joint splices [20] and (c) crack sizes from a few micrometres up to many millimetres.

An example is given in Fig. 4.1, which shows an exponential (or log–linear) presentation of QF-obtained FCG data for the lower wing skin of an F-111 FSFT wing [40] after completion of the test. The test article had also experienced the equivalent of 4500 h of service life before the FSFT. From these very extensive QF data, it could be determined that

i. The FCG behaviour was generally approximately exponential, and during the FSFT was very similar to that in service. Also, the FCGRs (i.e. slopes) were generally similar.

ii. In most cases, the cracks began to grow very early in the life of the wing.

iii. Cracks had grown at numerous locations covering two-thirds of the wingspan. This remarkably broad coverage demonstrates that when the stressing conditions are similar, the variations in geometrical detail and—in this case—spanwise location are of secondary importance. This is an important and useful result.

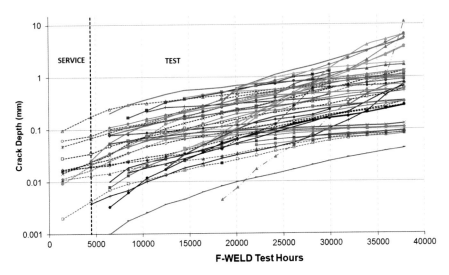

Fig. 4.1 A compilation of FCG curves from the AA2024-T8 lower wing skin (including **central spars (Green), forward auxiliary spar stations (Blue), pylons (Red)**) and lower D6AC pivot fitting (**Black**) of an F-111 test article, described in [40]. Note that each point represents the crack growth increment for one flight or test block (*created by author*)

iv. The effective sizes of the nucleating discontinuities varied significantly. This is also an important result.

These trends have also been observed for FSFTs of other types of aircraft [4, 25–32].

As stated in Chap. 2, there are significant advantages in using an exponential presentation of the data. Firstly, the early FCG behaviour is more evident than in a linear–linear presentation. Secondly, given the similar FCG rates, it can be seen that the major source of scatter in the total lives was the effective crack-like sizes of the nucleating discontinuities. These latter characteristics enabled a simplified LCFLF FCG curve to be derived for lifing purposes, as follows. Based on the expectation of almost immediate crack nucleation and exponential FCG behaviour, one can choose (i) an appropriate initial effective crack-like size together with the known or estimated final crack size (not necessarily the critical crack size) and (ii) an average exponent to directly obtain an FCG curve covering the known life. This curve can then be used to determine the FCG life at any other crack sizes, either by interpolation for intermediate crack sizes or extrapolation to any crack size up to the critical crack size i.e. as per the long-neglected Eq. 2.3.

The LCFLF supporting this simplified lifing method is considered to provide conservative life estimates [23], albeit that considerable engineering expertise is required for a judicious choice of the FCG curve parameters. The application of this framework for the interpretation of FSFT results is described in Chap. 7.

References

1. S. Barter, L. Molent, N. Goldsmith, R. Jones, An experimental evaluation of fatigue crack growth. Eng. Fail. Anal. **12**(1), 99–128 (2005)
2. D.L. Simpson, N. Landry, J. Roussel, L. Molent, A.D. Graham, N. Schmidt, The Canadian and Australian F/A-18 international follow-on structural test project, in *Proceedings ICAS 2002 Congress*, Toronto, Canada, September (2002)
3. L. Molent, S. Barter, P. White, B. Dixon, Damage tolerance demonstration testing for the Australian F/A-18. Int. J. Fatigue **31**, 1031–1038 (2009)
4. L. Molent, S.A. Barter, A comparison of crack growth behaviour in several full-scale airframe fatigue tests. Int. J. Fatigue **29**, 1090–1099 (2007)
5. R.J.H. Wanhill, Damage tolerance engineering property evaluations of aerospace aluminium alloys with emphasis on fatigue crack growth, NLR TP 94177, The Netherlands (1994)
6. P. White, S. Barter, L. Molent, Observations of crack path changes caused by periodic underloads in AA7050-T7451. Int. J. Fatigue **30**, 1267–1278 (2008)
7. S.A. Barter, L. Molent, R.J.H. Wanhill, Marker loads for quantitative fractography of fatigue cracks in aerospace alloys, in *Proceedings ICAF 2009*, Rotterdam, 27–29 May (2009)
8. R.J.H. Wanhill, N. Goldsmith, L. Molent, Quantitative fractography of fatigue and an illustrative case study. Eng. Fail. Anal. **19**, 426–435 (2019)
9. S. Barter, B. Dixon, L. Molent, Assessing relative severity using single fatigue test coupons. Eng. Fail. Anal. **16**(3), 863–873 (2009)
10. L. Molent, B. Aktepe, Review of fatigue monitoring of agile military aircraft. Fatigue Fract. Eng. Mater. Struct.Fract. Eng. Mater. Struct. **23**, 767–785 (2000)
11. L. Molent, J. Agius, Structural health monitoring of agile military aircraft, in *Encyclopedia of Structural Health Monitoring*, eds. by C. Boller, F. Chang, Y. Fujino, John Wiley & Sons, Ltd., (2008)
12. T. Dickinson, L. Molent, Validation of fatigue damage models used for F/A-18 life assessment using fatigue coupon test results, DSTO-TR-0940, Defence Science and Technology Organisation, Melbourne, Australia, February (2000)
13. P. White, D. Mongru, L. Molent, A crack growth based individual aircraft monitoring method utilizing a damage metric. Struct. Health Monit.. Health Monit. **17**(5), 1178–1191 (2018)
14. Methodology for determination of equivalent flight hours and approaches to communicate usage severity, USAF Structural Bulletin EN-SB-09-001, June (2006)
15. B. Aktepe, K.P. Hewitt, L. Molent, R.W. Ogden, Validation of RAAF F/A-18 analytical fatigue strain sensor calibration techniques using an aircraft ground calibration procedure, DSTO-TR-0641, Defence Science and Technology Organisation, Melbourne, Australia, January (2000)
16. Airworthiness Assurance Working Group for the Aviation Rulemaking Advisory Committee's Transport Aircraft and Engine Issues Group. Recommendations for Regulatory Action to Prevent Widespread Fatigue Damage in the Commercial Airplane Fleet. Final report, Revision A, 29 June (1999)
17. Aircraft Structural Integrity Program (ASIP), MIL-STANDARD-1530D, USA, August (2016)
18. Aloha Airlines, Flight 243, Boeing 737-200, N73711, Near Maui, Hawaii April 28, 1988, NTSB Report AAR-89-03 (1989)
19. L. Molent, R. Jones, Crack growth and repair of multi-site damage of fuselage lap joints. Eng. Fract. Mech.Fract. Mech. **44**(4), 627–637 (1993)
20. R. Jones, L. Molent, S. Pitt, Understanding crack growth in fuselage lap joints. Theor. Appl. Fract. Mech.. Appl. Fract. Mech. **49**, 38–50 (2008)
21. B. Dixon, L. Molent, S. Barter, A study of fatigue variability in aluminium alloy 7050–T7451. Int. J. Fatigue **92**(1), 130–146 (2016)
22. L. Molent, S.A. Barter, The lead fatigue crack concept for aircraft structural integrity. Proc. Eng. **2**, 363–377 (2010)
23. L. Molent, S.A. Barter, R.J.H. Wanhill, The lead crack fatigue lifing framework. Int. J. Fatigue **33**, 323–331 (2011)

24. L. Molent, S.A. Barter, R.J.H. Wanhill, The lead crack fatigue lifing framework, DSTO-RR-0353, Defence Science and Technology Organisation, Melbourne, Australia, April (2010)
25. L. Molent, S.A. Barter, R.J.H. Wanhill, The lead crack concept 30 years on, in *Proceedings ICAF 2023*, ed. by M. Bos, Delft, The Netherlands, 26–29 June (2023)
26. L. Molent, Fatigue crack growth from flaws in combat aircraft. Int. J. Fatigue 32, 639–649 (2010)
27. L. Molent, S.A. Barter, R. Jones, Some practical implications of exponential crack growth, in *Multiscale Fatigue Crack Initiation and Propagation of Engineering Materials: Structural Integrity and Microstructural Worthiness*, eds. by G.C. Sih, June 2008 (Springer Press, 2008). ISBN 978-1-4020-8519-2
28. L. Molent, A. Spagnoli, A. Carpinteri, R. Jones, Using the lead crack concept and fractal geometry for fatigue lifing of metallic structural components. Int. J. Fatigue 102C, 214–220 (2017)
29. S.A. Barter, L. Molent, R.J.H. Wanhill, Typical fatigue-initiating discontinuities in metallic aircraft structures. Int. J. Fatigue 41, 11–22 (2012)
30. L. Molent, A. Haddad, A critical review of available composite damage growth test data under fatigue loading and implications for aircraft sustainment. Compos. Struct.Struct. 232, 111568 (2020)
31. L. Molent, C. Forrester, The lead crack concept applied to defect growth in aircraft composite structures. Compos. Struct.Struct. 166, 22–26 (2017)
32. K. Maxfield, M. McCoy, D. Williams, R. Ogden, V.T. Mau, A. Zammit, in Failure analysis of a military transport aircraft fatigue test, in *Proceedings. Aircraft Structural Integrity Program (ASIP) 2018*, Phoenix, Arizona, USA (2018)
33. ASI-DGTA. F/A-18 A/B Structural Analysis Methodology (SAM), Issue 2 AL 3. Royal Australian Air Force, December (2007)
34. B. Main, L. Molent, R. Singh, S. Barter, Fatigue crack growth lessons from thirty-five years of the Royal Australian Air Force F/A-18 A/B Hornet aircraft structural integrity program. Int. J. Fatigue 133, 105426 (2020)
35. L. Molent, Q. Sun, A. Green, Characterisation of equivalent initial flaw sizes in 7050 aluminium alloy. Fatigue Fract. Eng. Mater. Struct.Fract. Eng. Mater. Struct. 29, 916–937 (2006)
36. L. Molent, A review of equivalent pre-crack sizes in aluminium alloy 7050–T7451. Fatigue Fract. Eng. Mater. Struct.Fract. Eng. Mater. Struct. 37, 1055–1074 (2014)
37. L. Molent, M. Fox, Crack-like effectiveness of some discontinuities in AA2024. Fat. Fract. Eng. Mat. Struc, 1–14 (2023)
38. B. Main, M. Jones, S. Barter, The practical need for short fatigue crack growth rate models. Int. J. Fatigue 142, 105980 (2021)
39. B. Main, M. Jones, B. Dixon, S. Barter, On small fatigue crack growth in AA7085-T7452. Int. J. Fatigue 156, 106704 (2022)
40. L. Molent, R. Boykett, K. Walker, Maintaining reliability and availability in a sole operator environment, RTO-MP-AVT-157, RTO Meeting, Montreal, Canada, 13–16 October (2008)

Chapter 5
Derivative Fatigue Crack Growth Models

Following on from the lead crack framework, a number of FCG tools have been developed:

- The cubic rule.
- The block-by-block (Dblock) or Effective Block Approach (EBA).
- The Hartman–Schijve FCG equation variant.

A brief description of each follows.

5.1 The Cubic Rule

The cubic rule represents a special category of lead crack exponential FCG, see Eq. (5.1) [1, 2]:

$$\text{FCG: } a = a_0 \exp\left(\sigma_{\text{REF}}^{\alpha} \lambda N\right) \tag{5.1}$$

where α is a constant, a_0 is the initial crack size and 'N' is the FCG life (number of load blocks, etc.). 'λ' is the crack growth rate (e.g. $da/dBlock$) which is material and spectrum dependent. We note here that for FSFTs the spectrum load histories are applied in repeated fixed blocks, each representing a period of service. From Eq. (5.1), the FCGR may be expressed as

$$\text{FCGR}: \; da/dN = a_0 \sigma_{\text{REF}}^{\alpha} \lambda \cdot \exp\left(\sigma_{\text{REF}}^{\alpha} \lambda N\right) = a \sigma_{\text{REF}}^{\alpha} \lambda \tag{5.2}$$

Equation (5.2) shows that the FCGR at any given crack length a is determined by $\sigma_{\text{REF}}^{\alpha} \lambda$. Then the FCGR ratio for two tests using the same load spectrum (same λ), but at two different reference stress levels, may be expressed as

© The Author(s), under exclusive license to Springer Nature Singapore Pte Ltd. 2024
L. Molent, *Aircraft Fatigue Management*, SpringerBriefs in Applied Sciences
and Technology, https://doi.org/10.1007/978-981-99-7468-9_5

$$(da/dN)_1/(da/dN)_2 = (\sigma_{REF.1}/\sigma_{REF.2})^{\alpha} \qquad (5.3)$$

Equation (5.3) implies that for a given load spectrum the lead FCGR obtained at one reference stress level may be used to predict the lead FCGR for a different reference stress level. This equation is very useful because tests with several load spectra and different aerospace structural alloys, including aluminium alloys, a titanium alloy and a high-strength steel, have shown that $\alpha \approx 3$ [2]. Owing to this result, Eq. (5.3) with $\alpha \approx 3$ has been designated the cubic (stress-cubed) rule.

Practical applications of the cubic rule include life predictions for structural repairs [3], and the use of QF-derived lead FCGR data from one location and material to predict the lead crack behaviour at other locations that experience the same load spectrum but at different stress levels.

5.2   The Block-By-Block EBA Approach

The spectrum block-by-block (dBlock) EBA approach [1, 4–8] is a framework whose purpose is to predict the FCG lives of in-service aircraft structures subjected to relatively short and repeated blocks of VA loading sequences, i.e. load sequences representative of tactical (combat) aircraft. The EBA has a long history, dating from the mini-block proposal by Gallagher in 1976 [9]. The EBA has been developed using extensive experience with VA FSFTs in combination with specimen tests and QF. This combination was essential in re-lifing the RAAF's F-111 wings [6, 10].

The EBA approach treats a block of VA FCG as if it were effectively CA data, based partly on the analysis of FCG test data obtained from VA block loading sequences simulating those experienced by tactical aircraft. Use of the EBA for predicting service-induced (i.e. untested) FCG requires several additional inputs and conditions/assumptions:

(1)   QF of fatigue fracture surfaces from full-scale or specimen tests (see point 2) to find FCG progression markings and measure the corresponding crack sizes. This is to obtain crack size, a, versus NB data, where NB is the number of applied VA blocks.
(2)   The FCG test data must be for the same material and for the same location for which a prediction is required for a VA block loading history (as many repeat blocks as possible) considered to be representative of typical in-service usage.
(3)   Each block of VA loading may be treated as if it were a single CA cycle, and the FCG associated with each block should correlate well with a similitude parameter, e.g. an appropriate (characteristic) K-value.
(4)   QF data from specimen tests or FSFT (or both) may be used, but only one source of test data is actually required.
(5)   Conversion and presentation of the QF FCG data according to a da/dN_B versus characteristic K relationship:

$$da/dN_B = C_{va}(K_{REF})^{m_{va}} \qquad (5.4)$$

where C_{VA} and m_{VA} are analogous to the constant and exponent in the Paris equation, see Chap. 2, Sect. 2.2; and K_{REF} is the reference (characteristic) K-value given by $K_{REF} = \sigma_{REF}\beta\sqrt{\pi a}$, where σ_{REF} is a reference stress (e.g. peak stress) from the VA block loading sequence.

The C_{VA} for the untested spectrum is determined via use of an additional LEFM-based FCG model, for example, AFGROW, see [7]. However, for both spectra m_{VA} is set $= 2$ (i.e. exponential growth), based on the results of many EBA tests. A representative example is given in Fig. 5.1, which shows converted QF test data, i.e. FCGR, plotted against K_{REF} using log–log coordinates. It is seen that m_{VA} provides good fits to the data. N.B.: This result (and many others) is highly significant since, as already pointed out in Sect. 2.2, an exponent of 2 corresponds to exponential FCG, i.e. a straight line when the log crack size (depth) is plotted against linear life N (in this case N_B).

Similar observations were made by Davidson and Lankford [11]: 'In examining fatigue crack growth rate curves for many materials exhibiting very large differences in microstructure, the striking feature is the similarities between these curves, not the differences'.

Thus in practice, if QF data are available for one spectrum and one σ_{REF} for a given material, then a prediction (or lifing) can be made for an untested spectrum, such as another in-service spectrum, at another stress level for that same material. In addition, a typical equivalent pre-crack size (EPS) (see Chap. 6, Sect. 6.2) and a_{CRIT} are required for the analysis.

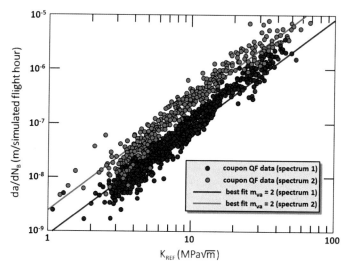

Fig. 5.1 FCGR data for two EBA block loading spectra and the best fits with $m_{va} = 2$. The QF data were obtained from many AA7050-T7451 coupons tested at four σ_{REF} levels per spectrum: after [5, 7], and courtesy of R. J. H. Wanhill

5.3   The Modified Hartman–Schijve Equation

Extending an equation proposed by Hartman and Schijve [12], the following empirical relation was developed [13–16]:

$$\frac{da}{dN} = \left(\frac{D(\Delta K - \Delta K_{\text{thr}})}{\sqrt{\left(1 - \frac{K_{\text{max}}}{A}\right)}} \right)^{\alpha} \tag{5.5}$$

where K_{max} is the maximum stress intensity, D is the da/dN versus ΔK y-axis intercept at approximately 1 MPa $\sqrt{m}$, A is a 'fracture toughness-equivalent' parameter and α is approximately 2. Only one set of da/dN versus ΔK data at one stress ratio is required to define the material-dependent parameter D.

Equation (5.5) was developed to allow predictions of spectrum fatigue lives. Following the observation that ΔK_{THRS} is very low for short cracks, this parameter can conservatively be set to a low value. Effectively this means that Eq. (5.5) becomes a two-parameter equation, in which A is chosen to allow the best fit to the available FCGR data. The usefulness of this equation has been demonstrated for a complex spectrum and geometry in Ref. [17] and is included in the DST Easigro program [18]. It was noted in [16] that this is a special case of the well-known NASGROW equation.

References

1. S. Barter, L. Molent, N. Goldsmith, R. Jones, An experimental evaluation of fatigue crack growth. Eng. Fail. Anal. **12**(1), 99–128 (2005)
2. L. Molent, R. Jones, A stress versus crack growth rate investigation (aka stress – cubed rule). Int. J. Fatigue **87**, 435–443 (2016)
3. J. Ayling, G. Bowler, M. Brick, M. Ignjatovic, Practical application of structural repair fatigue life determination on the AP-3C Orion platform. Adv. Mat. Res. **891–892**, 1065–1070 (2014)
4. L. Molent, M. McDonald, S. Barter, R. Jones, Evaluation of spectrum fatigue crack growth using variable amplitude data. Int. J. Fatigue **30**(1), 119–137 (2007)
5. L. Molent, M. McDonald, S. Barter, R. Jones, Evaluation of spectrum fatigue crack growth using variable amplitude data. Int. J. Fatigue **30**, 119–137 (2008)
6. W.Z. Zhuang, M. McDonald, M. Phillips, L. Molent, Effective block approach for aircraft damage tolerance analyses. J. Aircr.Aircr **46**(5), 1660–1666 (2009)
7. M. McDonald, Guide on the effective block approach for fatigue life assessment of metallic structures, DSTO Technical Report DSTO-TR-2850, Defence Science and Technology Organisation, Melbourne, Australia (2013)
8. W. Zhuang, R. Boykett, M. Phillips, L. Molent, Effective block approach for damage tolerance analyses of F-111 D/F model wing structures, DSTO-TR-2124, Defence Science and Technology Organisation, Melbourne, Australia, May (2008)
9. J.P. Gallagher, Estimating fatigue-crack lives for aircraft: techniques. Exp. Mech. **16**(11), 425–433 (1976)
10. L. Molent, R. Boykett, K. Walker, Maintaining reliability and availability in a sole operator environment, RTO-MP-AVT-157, RTO Meeting, Montreal, Canada, 13–16 October (2008)

11. D.L. Davidson, J. Lankford, Fatigue crack growth in metals and alloys: mechanisms and micromechanics. Int. Mater. Rev. **37**(2), 45–75 (1992)
12. A. Hartman, J. Schijve, The effects of environment and load frequency on the crack propagation law for macro fatigue crack growth in aluminum alloys. Eng. Fract. Mech.Fract. Mech. **1**(4), 615–631 (1970)
13. R. Jones, L. Molent, K. Walker, Fatigue crack growth in a diverse range of materials. Int. J. Fatigue **40**, 43–50 (2012)
14. , R. Jones, L. Molent, S. Barter, Calculating crack growth from small discontinuities in 7050-T7451 under combat aircraft spectra. Int. J. Fatigue **55**, 178–182 (2013)
15. L. Molent, R. Jones, The influence of cyclic stress intensity threshold on fatigue life scatter. Int. J. Fatigue **82**, 748–756 (2015)
16. R. Jones, Fatigue crack growth and damage tolerance. Fatigue Fract. Eng. Mater. Struct.Fract. Eng. Mater. Struct. **37**, 463–483 (2014)
17. B. Main, R. Evans, K. Walker, X. Yu, L. Molent, Lessons from a fatigue prediction challenge for an aircraft wing shear tie post. Int. J Fatigue **123**, 53–65 (2019)
18. P. White, A guide to the program Easigro for generating optimised fatigue crack growth models. DST-Group-TR- 3566, Defence Science and Technology Group, Melbourne, Australia, February (2019)

Chapter 6
Fatigue-Nucleating Discontinuities

Compared to the carefully prepared surfaces typical of traditional laboratory fatigue specimens, production aircraft structures have many surface discontinuities. These discontinuities are generally very small, of the order of 0.01 mm deep. However, when they are sufficiently crack-like and the local cyclic stresses generated by the service loads are sufficiently high, the nucleation and growth of fatigue cracks will commence very early—almost immediately—in an aircraft's service life or in an FSFT [1–11]. This almost immediate growth of cracks from such small discontinuities has also been noted by others, e.g. [12–14]. In fact, while not explicitly stated, this assumption is inherent to the USAF and FAA damage-tolerant design methods.

The discontinuities include [15–17]

a. Machining damage: in the form of grooves, small surface tears and nicks, scratches from the removal of fasteners (drills, reamers, etc.), burrs and poorly de-burred holes, fretting.
b. Inherent material discontinuities: broken intermetallic particles at or just below machined or abraded surfaces.
c. Porosity at or just below the surface.
d. Surface treatment damage: anodizing or etch pitting from chemical treatments; laps, folds and embedded particles from shot peening and grit blasting.
e. Time-dependent discontinuities such as corrosion pits, wear, and in-service-induced damage.

Examples of likely discontinuities, including two that could be classified as very small 'defects' (if avoidable in manufacturing) are shown in Fig. 6.1.

A possible source of an apparent nucleation period is FCG from sub-surface discontinuities that effectively begin to grow in vacuo. However, sub-surface discontinuities are rarely the critical nucleating ones (possibly excluding highly polished, cold-worked and burnished surfaces, notably for engine components and hence beyond the scope of this book), since it is well known that FCG in vacuo is much slower than FCG for an equivalent surface-breaking crack growing in air [18]. N.B.

L. Molent, *Aircraft Fatigue Management*, SpringerBriefs in Applied Sciences and Technology, https://doi.org/10.1007/978-981-99-7468-9_6

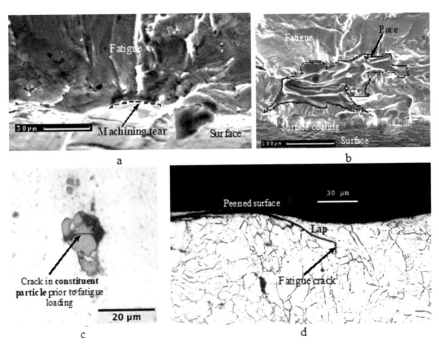

Fig. 6.1 Examples of discontinuities that can (and do) exist in metallic airframes when they enter service. Each discontinuity will act as an effective crack-starter that under sufficiently high local cyclic stresses will reduce fatigue nucleation to a few load cycles (*source* author and colleagues)

It should be noted that once a sub-surface crack breaks through to the material/component's surface the FCGR soon becomes similar to the FCGRs of similarly located cracks that had nucleated from surface-breaking discontinuities, see Fig. 2. 3 in Chap. 2 and Ref. [19].

The risk of an in-service fatigue failure may be considered as a 'race to the critical crack size' by cracks that almost immediately nucleate at or near the surface from pre-service discontinuities, and cracks that nucleate from service-induced damage (including corrosion) that occur at some time during the service life. This 'race' assumes the absence of unusually large or 'rogue' flaws. As noted in Ref. [6], undetected rogue flaws lead to fatigue failures early in the life of type (i.e. the extremes of a population are revealed early if a rogue flaw exists).

Corrosion pit-nucleated FCG has also been shown to be approximately exponential [20–23]. Importantly, it has been shown that for many aircraft FCG is decoupled from environmental exposure to (possible) corrosion, i.e. fatigue occurs in flight while corrosion occurs during static stays on the ground [20–23]. This is especially important for military aircraft, whose periods on the ground usually are measured in many days or several weeks, unlike the high utilization of commercial aircraft.

6.1 Discontinuity Sources

QF data for a large number of F/A-18 Hornet FSFTs were studied to determine the features and production aspects leading to fatigue cracking [16, 17, 24]. These data are summarized in Fig. 6.2, which shows that most of the fatigue cracks grew from etch pits (This population may be biased, e.g. against corrosion pits, because some FSFT articles did not see service, and many holes contained neat fit fasteners). The etch pits resulted from chemical pre-treatment of aluminium alloy airframe components before applying an Ion Vapour Deposition (IVD) aluminium coating to improve the corrosion resistance. Despite the predominance of etch pits, there were other fatigue-nucleating features such as near-surface porosity, constituent particles, mechanical damage and corrosion pits.

Most of the mechanical damage, which was a significant contributor (9%) to the overall fatigue-cracking problem, occurred in the bores of fastener holes. There were also a few instances of poor shot peening causing fatigue cracking. However, these cracks grew more slowly than similar cracks nucleating from non-peened surfaces. This is unsurprising, since shot peening is normally a life-enhancement process, introducing surface and near-surface compressive stresses.

Finally, there were many unidentifiable sites (about 21% of all sites) where FCG commenced. These sites could not be identified because the crack origins were obliterated by subsequent damage during the FSFTs. Taken literally in the round, the results shown in Fig. 6.2 highlight the importance of surface finish and production quality on the fatigue lives of aircraft structures. While this observation certainly is not new, e.g. see Ref. [25], it seems to have been forgotten by some researchers.

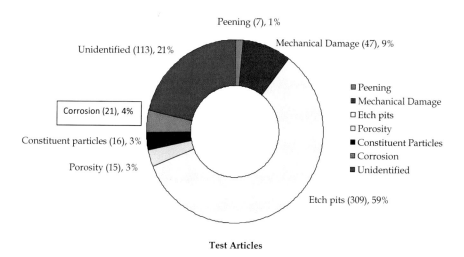

Test Articles

Fig. 6.2 Percentages (and number) of FCG-nucleating discontinuities found in F/A-18 FSFT articles [24] (*created by author*)

6.1.1 Equivalent Pre-crack Size (EPS)

Under otherwise identical conditions, variation in the effective crack-like size of the initial discontinuities, see Sect. 3.9, is an important factor in determining the fatigue life and its variability. The effective crack-like size discussed here has been called the EPS to distinguish it from the United States Air Force (USAF) analytical Equivalent Initial Flaw Size (EIFS) or Equivalent Initial Discontinuity State (EIDS) concepts: see, for example, Refs. [14, 16, 26].

A relatively simple method has been used to estimate EPS values. This method was, and is, based on QF analyses of many in-service and FSFT lead cracks in AA 7050-T7451, as well as in many other investigated materials, see, for example, Fig. 6.3. These analyses showed that the lead cracks grew in an approximately exponential fashion, as discussed in Chap. 4, and they also began to grow almost immediately, also mentioned in Chap. 4.

The general method used to define the EPS requires the fitting of an exponential equation, i.e. Eq. (2.3) in Chap. 2, Sect. 2.2, to the QF-determined FCG data. In this fit, 'a_0' is the initial apparent crack depth/length, i.e. the EPS. The a_0 value, and hence the EPS, will depend on which curve-fitting technique is used and the selection of fitted data points. This is an instance of analytical treatment of real data that needs much engineering expertise and judgement.

To understand more about the EPS, in particular, the values resulting from fatigue crack nucleation at etch pits, we refer to Chap. 7, Sect. 7.1, concerning an investigation involving a large number of coupon tests. The coupons were made from the same AA7050 as in the F/A-18 Hornet FSFTs and were subjected to the same chemical pre-treatment and IVD process [16]. Figure 6.3 presents a selection of EPS data

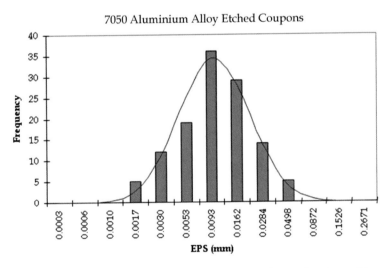

Fig. 6.3 Log-normal EPS distribution from etched coupons—120 test values (*created by author*)

Fig. 6.4 Log-normal distribution of the 71 AA2024 EPS values [27] (*source* author and colleague)

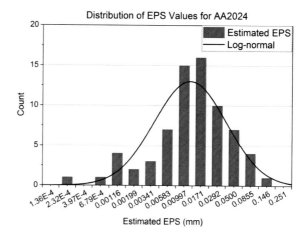

from this coupon test investigation. The data fit well to a log-normal distribution, noting that there were few data in the tails to confirm this. The average EPS is approximately 0.01 mm deep, which is partly the reason this value was mentioned under generalization (4) in Chap. 4, Sect. 4.1. The usefulness of assuming an EPS value of 0.01 mm for a lead FCG analysis is demonstrated in Chap. 7.

A similar assessment was also conducted for aircraft production quality AA2024, tested under multiple aircraft spectra, using the lead crack concept [27], with the results presented in Fig. 6.4. The data fit well to a log-normal distribution, noting that there were few data in the tails to confirm this. The average EPS depth is again approximately 0.01 mm.

References

1. L. Molent, S.A. Barter, A comparison of crack growth behaviour in several full-scale airframe fatigue tests. Int. J. Fatigue **29**, 1090–1099 (2007)
2. R.J.H. Wanhill, N. Goldsmith, L. Molent, Quantitative fractography of fatigue and an illustrative case study. Eng. Fail. Anal. **19**, 426–435 (2007)
3. R. Jones, L. Molent, S. Pitt, Understanding crack growth in fuselage lap joints. Theor. Appl. Fract. Mech.. Appl. Fract. Mech. **49**, 38–50 (2008)
4. L. Molent, S.A. Barter, R.J.H. Wanhill, The lead crack fatigue lifing framework. Int. J. Fatigue **33**, 323–331 (2011)
5. L. Molent, S.A. Barter, R.J.H. Wanhill, The lead crack concept 30 years on, in *Proceedings ICAF 2023*, ed. by M. Bos, Delft, The Netherlands, 26–29 June (2023)
6. L. Molent, Fatigue crack growth from flaws in combat aircraft. Int. J. Fatigue **32**, 639–649 (2010)
7. L. Molent, S.A. Barter, R. Jones, Some practical implications of exponential crack growth, in *Multiscale Fatigue Crack Initiation and Propagation of Engineering Materials: Structural Integrity and Microstructural Worthiness*, eds. by G.C. Sih, June 2008 (Springer Press, 2008). ISBN 978-1-4020-8519-2

8. L. Molent, A. Spagnoli, A. Carpinteri, R. Jones, Using the lead crack concept and fractal geometry for fatigue lifing of metallic structural components. Int. J. Fatigue **102C**, 214–220 (2017)
9. K. Maxfield, M. McCoy, D. Williams, R. Ogden, V.T. Mau, A. Zammit, Failure analysis of a military transport aircraft fatigue test, in *Proceedings. Aircraft Structural Integrity Program (ASIP) 2018*, Phoenix, Arizona, USA (2018)
10. B. Main, L. Molent, R. Singh, S. Barter, Fatigue crack growth lessons from thirty-five years of the Royal Australian Air Force F/A-18 A/B Hornet aircraft structural integrity program. Int. J. Fatigue **133**, 105426 (2020)
11. L. Molent, R. Boykett, K. Walker, Maintaining reliability and availability in a sole operator environment, RTO-MP-AVT-157, RTO Meeting, Montreal, Canada, 13–16 October (2008)
12. Y. Murakami, K.J. Miller, What is fatigue damage? A view point from the observation of low cycle fatigue process. Int. J. Fatigue **27**(8), 991–1005 (2005)
13. N. Thompson, Experiments relating to the basic mechanism of fatigue, in *International Conference on Fatigue of Metals, 1956, Proceedings,* Institute of. Mechanical Engineers, London, (1957)
14. A.P. Berens, P.W. Hovey, D.A. Skinn, Risk analysis for aging aircraft fleets. Analysis, vol 1, USAF WL-TR-91-3066, USA (1991)
15. S.A. Barter, L. Molent, R.J.H. Wanhill, Typical fatigue-initiating discontinuities in metallic aircraft structures, Int. J. Fatigue **41**, 11–22 (2012)
16. L. Molent, Q. Sun, A. Green, Characterisation of equivalent initial flaw sizes in 7050 aluminium alloy. Fatigue Fract. Eng. Mater. Struct.Fract. Eng. Mater. Struct. **29**, 916–937 (2006)
17. L. Molent, A review of equivalent pre-crack sizes in aluminium alloy 7050–T7451. Fatigue Fract. Eng. Mater. Struct.Fract. Eng. Mater. Struct. **37**, 1055–1074 (2014)
18. R.J.H. Wanhill, Fractography of fatigue crack propagation in 2024–T3 and 7075–T6 aluminum alloys in air and vacuum. Metall. Trans. A **6**(8), 1587–1596 (1975)
19. L. Molent, S.A. Barter, R.J.H. Wanhill, The lead crack fatigue lifing framework, DSTO-RR-0353, Defence Science and Technology Organisation, Melbourne, Australia, April (2010)
20. L. Molent, S. Barter, R.J.H. Wanhill, The decoupling of corrosion and fatigue for aircraft service life management, in *Proceedings ICAF Symposium*, Helsinki, 3–5 June (2015)
21. L. Molent, Managing airframe fatigue from corrosion pits – a proposal. Eng. Fract. Mech.Fract. Mech. **137**, 12–25 (2015)
22. S.A. Barter, L. Molent, Fatigue cracking from a corrosion pit in an aircraft bulkhead. Eng. Fail. Anal. **39**, 155–163 (2014)
23. L. Molent, R.J.H. Wanhill, Management of airframe in-service pitting corrosion by decoupling fatigue and environment. Corros. Mater. Degrad. **2**, 493–511 (2021)
24. L. Molent, Q. Sun, The Compendium of F/A-18 Hornet Crack Growth Data Version II: Part 1. DSTO-TR-2488, Defence Science and Technology Organisation, Melbourne, Australia, November (2010)
25. Prevention of failure of metals under repeated stress. Battelle Memorial Institute, (Wiley and Sons, New York, 1941)
26. J.P. Gallagher, L. Molent, The equivalence of EPS and EIFS based on the same crack growth life data. Int. J. Fatigue, **80**, 162–170 (2015)
27. L. Molent, M.R. Fox, Crack-like effectiveness of some discontinuities in AA2024. Fract. Eng. Mater. Struct. 1–14 (2023)

Chapter 7
Lead Crack Lifing Methodology

The lead FCG concept has been developed into a methodology that can be used to estimate virtual test endpoints. These test endpoints can be obtained from FSFTs and large component fatigue tests; in-service fleet cracking, including disassembled (teardown) parts taken from the fleet and sometimes from highly representative coupon tests. The methodology may be separated into three distinct assessments:

1. Fatigue crack nucleation and the EPS of the initial discontinuities and defects.
2. The FCG behaviour, notably determination of the parameter λ. As noted in Chap. 5, Sect. 5.1, λ is material and spectrum dependent. If the FCG is approximately exponential, λ may be used to define a partial or overall exponential crack growth rate, whatever the load history and spectrum type.
3. The RST critical crack size for 1.2 X DLL (either demonstrated or analytically obtained).

Combining these three assessments with the assumption of immediate FCG from the beginning of cyclic loading enables virtual test endpoints to be determined for locations that do not have a demonstrated life up to the time at which the RST is applied without failure.

An example showing the practical benefit of this methodology is given in Fig. 7.1. This figure shows an exponential FCG curve determined via QF for a test article location that was repaired well before it would have failed under 1.2 X DLL. The virtual test endpoint (equivalent RS life) for the unrepaired location is obtained by a simple extrapolation to the critical crack depth. This method has also been suggested as a potential way of reducing the amount of required cycling for an FSFT, to expedite the delivery of durability results to the fleet manager [1].

As mentioned earlier, although this example appears straightforward, the choice of the parameters used in this methodology requires considerable engineering expertise and judgement. For example, in this case, the critical crack size, acrit, was conservatively chosen as the thickness of the component's web (thinner section), since it

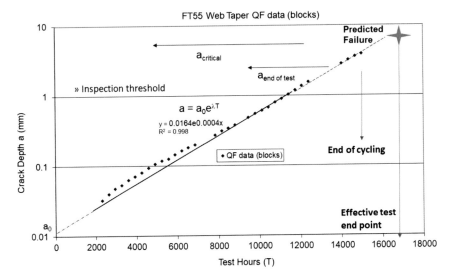

Fig. 7.1 QF-determined FCG of a typical crack (from [2]) from an F/A-18 FSFT showing the crack depth versus time history (test hours T) for blocks of spectrum loading. The QF data are back-extrapolated to an initial crack size based on the assumption of immediate FCG, and forward-extrapolated to a conservatively estimated critical crack size. The solid line can be considered a prediction (*created by author*)

was considered that the FCGR would have increased (by an unknown amount) once the crack had penetrated into the web.

There are cases with only partially exponential FCG [3–6], e.g. when load shedding occurs as a crack grows. It is considered—again based on acquired expertise—that the LCFLF methodology can be used for some of these cases by careful choice of an exponential FCGR [3].

Although the LCFLF described above applies to spectrum test results, methods of adapting the observations of exponential FCG to predict the lives of fatigue specimens under yet-to-be-tested VA loading have also been developed. These methods, also based on the lead crack concept, have been summarized in Chap. 5.

7.1   Comparison of Aircraft Design Standards

A long-running and important debate in the aircraft industry relates to the relative efficiencies of airframes constructed according to different design standards. The assumption of exponential FCG behaviour for lead cracks enables a simple visual comparison of some of the current military airworthiness design regulations (limited here to fatigue life requirements), as will be shown below. Firstly, a brief summary of the requirements is given here:

DEFSTAN 00-970 [7] suggests that for an aircraft designed and managed by the safe-life method, an FSFT needs to be conducted for 5 or more times the expected service life of the airframe. This is based on a scatter factor of $3\,^1/_3$ to account for material fatigue performance scatter, and a further factor of 1.5 for structural elements that are not load-monitored in service.

USAF JSSG 2006 [8]. The traditional USAF part of this standard employs durability and damage tolerance (DT) analyses, underpinned by durability FSFT. The USAF durability FCG analyses require that an assumed manufacturing flaw (taken to be 0.254 mm deep) shall not grow to cause airframe functional impairment (here assumed to be a crack approximately 10 mm deep) in 2X the design service life. This requirement is to be demonstrated by 2 lifetimes of durability testing.

The USAF **DT** testing requires that an assumed manufacturing flaw (taken to be 1.27 mm deep) shall not grow to cause airframe functional impairment (here assumed to be a crack approximately 10 mm deep) in 1 X the design service life. This requirement can be demonstrated by one lifetime of durability testing. Additionally, if an analytical hotspot has not shown evidence of cracking, then the DT for such a location is usually (if possible, but not necessarily) validated on the same FSFT by introducing flaws (typically 1.27 mm semi-circular 'cracks') and testing for an additional one lifetime.

USN JSSG 2006 [8]. The traditional US Navy part of this standard requires that crack 'initiation' (an engineering term defined as a 0.254-mm-deep crack) should not occur during 2 lifetimes of durability testing from the natural flaw population, i.e. the airframe condition when the aircraft first enters service.

All of these four main standards' requirements can be represented in terms of exponential FCG in critical structures, starting from required induced/assumed flaw depths or from a typical crack-like discontinuity of 0.01 mm depth: this latter configuration was discussed already in Chaps. 4 and 8. The following additional assumptions have been made to allow for comparisons.

The critical structures have been well designed (optimized) such that at the end of the required FSFT cycling (regardless of the spectrum used) the cracked structure just meets the DEFSTAN residual strength requirement of 1.2 X DLL, or the USAF structural impairment criterion, or the USN crack 'initiation' criterion. Furthermore, it is assumed that

- The FSFT article is initially pristine (e.g. no exposure to a detrimental environment).
- The USAF failure or functional impairment criterion is equivalent to a 10-mm-deep crack.
- A 1.27-mm-deep crack-like flaw was induced/assumed for a crack-free hotspot at the end of 2 lifetimes of USAF durability testing.
- The design life value is the same in each case.

These assumptions enable the different design standards to be compared as shown in Fig. 7.2. It can be seen that

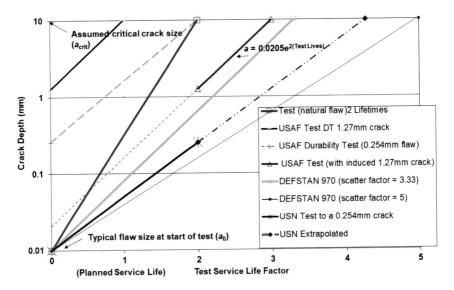

Fig. 7.2 Comparison of various aircraft fatigue design and test requirements, assuming a final crack length of 10 mm and a naturally occurring typical initial crack length of 0.01 mm (modified from [9], source author)

(1) An FSFT with a natural flaw with a two-lifetime requirement produces the fastest FCGR.

(2) Optimization of each of the standards leads to some differences in FCGRs to demonstrate one planned service life. However, it is remarkable that the FCG slopes for the USAF durability and DT requirements, the USN requirements and DEFSTAN 00-970 (scatter factor of $3\,^{1}/_{3}$) requirements are similar. This may be fortuitous, since these design standards are not formally related. However, each aims at achieving an equivalent POF. Notwithstanding this similarity, use of these different design standards will result in airframe structures with differing tolerance to fatigue cracking. The more conservative standards could lead to unnecessary structural weight penalties.

(3) Inducing/assuming a flaw of 1.27 mm depth at the end of 2 test lives is equivalent to growth from a naturally occurring 0.0205 mm flaw from zero hours (using back-extrapolation).

(4) Irrespective of the design standard and owing to the required POF, fleet structures should not have detectable (>1.0 mm) cracks present at the end of one lifetime when cracks grow from naturally occurring flaws of approximately 0.01 mm. However, for the USAF durability condition FCG could theoretically be possible after only one lifetime if the flaw was present and the fleet aircraft flew the 90th percentile (severe) spectrum.

Finally, it should be noted that at the end of any FSFT (usually no more than about three service lives) there will be small cracks in critical parts of the airframe, and a teardown inspection is (or would be) necessary to find them. Despite satisfactory/

favourable FSFT results, such small cracks could become critical during service owing, for example, to (i) insufficiently demanding FSFT conditions and/or more severe service usage than assumed; (ii) larger fatigue-nucleating discontinuities in service aircraft (possible, but unlikely) and (iii) aircraft weight increases (very likely).

7.2 In-Service Detected Cracks

As has been mentioned, unanticipated fatigue cracking does sometimes occur. When it occurs in primary structures and components, it is normal to ground the fleet until operational safety can be re-established. In general, the approaches available to the fleet manager are

i. Develop and employ a recurring inspection method (see below). Using advanced in situ Non-Destructive Inspection (NDI) sensors may be an option (see [10]).
ii. Replace the components or develop a more fatigue-tolerant design.
iii. In situ repair of the components.
iv. Determine the fleet POF if no action is taken, see Sect. 7.3.

Small cracks will always be present in highly loaded aircraft structures. In a well-designed aircraft, these cracks will—in theory—remain below the NDI threshold for at least one lifetime in the absence of unusually large discontinuities, see [10] and Fig. 7.2. Nevertheless, for reasons presented earlier, experience has shown that unexpected cracking is sometimes found in service aircraft (see, for example, Refs. [10–16]). Usually these cracks are detected using conventional NDI, including visual inspection, before they become critical. However, such incidents also cause significant concerns about safety, logistics and aircraft availability for airworthiness managers.

Once an unanticipated incidence of cracking is identified, it normally takes several years before a permanent logistical solution can be implemented, e.g. component redesign and replacement. In the interim, the airworthiness manager requires an inspection interval to maintain safety. As was first suggested in [17], the lead crack concept provides one basis for doing so. To illustrate this, Fig. 7.3 provides a schematic example. Suppose an in-service crack is detected at 5000 equivalent service hours. If a typical EPS of 0.01 mm is used as the starting point, and the size of the detected crack is known (here given as 1 mm), then the lead crack concept will provide an estimate of the effective life of the component. If many aircraft have been inspected, then these results can be pooled to enable and provide a better estimate of the average life. The steepest estimated FCG slope can be used to set a conservative inspection interval. N.B. The flight history of the aircraft is not required for this assessment: this is a significant advantage.

The lead crack concept has also been used to obtain the lives of crack-prone locations in service aircraft, based on the cracks found via the testing and teardown of ex-service components [18].

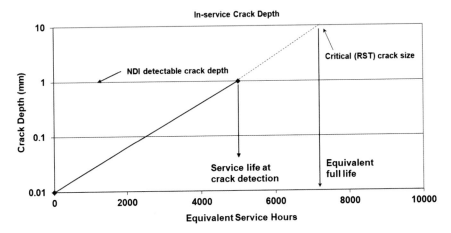

Fig. 7.3 Schematic demonstration of estimating the equivalent service hours to failure for in-service detected cracking (*created by author*)

7.3 Probability of Failure

An alternative to repair (or otherwise replacement) is to calculate the latent POF of an aircraft within the fleet and allow the fleet manager to determine whether the increased level of risk is acceptable. A single flight POF level of 1×10^{-7} is generally considered acceptable, while action should be taken if the POF increases above 1×10^{-5} [19, 20]. A novel method of performing the POF calculation, based on the lead FCG framework, is outlined in Ref. [20], which considers the variabilities in the primary uncertainties that govern the probability distribution of fatigue failure for an aircraft. These uncertainties are

a. EPS at the start of the airframe's fatigue life (i.e. based on estimated EPS distributions, see Chap. 8.)
b. FCGR.
c. Fracture toughness of the material.
d. Maximum load in each flight.
e. Accuracy in calculating the amount of fatigue life consumed, i.e. the amount of FCG resulting from a measured load history.

   The POF calculation relies on the lead crack concept for the FCG analysis and fitted distributions to the variables. In a reaction to this development, the United States Navy (USN) stated [21]:

   'If the initial crack distribution and the crack evolution law are accurately determined (from the experimental data and microstructural analysis), the risk of failure can be estimated by well-developed procedures' as detailed in Ref. [20] (Fig. 7.4).

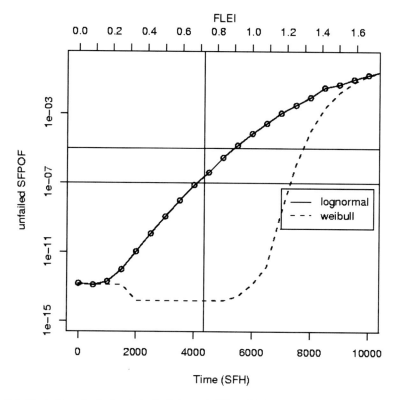

Fig. 7.4 Illustration of calculated single flight probability of failure showing the sensitivity to the assumed distribution of initial discontinuities (*created by author*)

References

1. L. Molent, R. Singh, Using the lead crack framework to reduce durability test duration. Aeronaut. J. **124**(1276), 814–820 (2020)
2. S.A. Barter, Fractographic investigation report FFT55 0030, Defence Science and Technology Organisation, Melbourne, Australia, February (2004)
3. L. Molent, S.A. Barter, R.J.H. Wanhill, The lead crack fatigue lifing framework. Int. J. Fatigue **33**, 323–331 (2011)
4. L. Molent, S.A. Barter, R.J.H. Wanhill, The lead crack fatigue lifing framework, DSTO-RR-0353, Defence Science and Technology Organisation, Melbourne, Australia, April (2010)
5. L. Molent, S.A. Barter, R.J.H. Wanhill, The lead crack concept 30 years on, in *Proceedings ICAF 2023*, eds. M. Bos, Delft, The Netherlands, 26–29 June (2023)
6. B. Main, L. Molent, R. Singh, S. Barter, Fatigue crack growth lessons from thirty-five years of the Royal Australian Air Force F/A-18 A/B Hornet aircraft structural integrity program. Int. J. Fatigue **133**, 105426 (2020)
7. Ministry of Defence, Defence Standard 00-970 Issue 1, Design and Airworthiness Requirements for Service Aircraft. Aeroplanes, vol 1, Amendment 13, December (1994)
8. Aircraft Structures, Joint Service Specification Guide JSSG-2006, US Department of Defense (1998)

9. L. Molent, S.A. Barter, A comparison of crack growth behaviour in several full-scale airframe fatigue tests. Int. J. Fatigue **29**, 1090–1099 (2007)
10. L. Molent, Considering the role of health monitoring in fixed wing airframe HUMS, Proceedings. HUMS 2003, Melbourne, Australia, 17–18 February (2003)
11. C.F. Tiffany, J.P. Gallagher, C.A. Babish IV, Threats to aircraft structural safety, including a compendium of selected structural accidents/incidents. ASC-TR-2010-5002, AFBWP, OH, USA, March (2010)
12. R.J.H. Wanhill, L. Molent, S.A. Barter, Milestone case histories in aircraft structural integrity, in *Comprehensive Structural Integrity*, Second edn. Elsevier Inc., (2023)
13. R.J.H. Wanhill, N. Goldsmith, L. Molent, Quantitative fractography of fatigue and an illustrative case study. Eng. Fail. Anal. **19**, 426–435 (2019)
14. R. Jones, L. Molent, S. Pitt, Understanding crack growth in fuselage lap joints. Theor. Appl. Fract. Mech.. Appl. Fract. Mech. **49**, 38–50 (2008)
15. L. Molent, Fatigue crack growth from flaws in combat aircraft. Int. J. Fatigue **32**, 639–649 (2010)
16. K. Maxfield, M. McCoy, D. Williams, R. Ogden, V.T. Mau, A. Zammit, Failure analysis of a military transport aircraft fatigue test, in *Proceedings Aircraft Structural Integrity Program (ASIP) 2018*, Phoenix, Arizona, USA (2018)
17. L. Molent, R. Singh, J. Woolsey, A method for evaluation of in-service fatigue cracks. Eng. Fail. Anal. **12**(1), 13–24 (2004)
18. S.A. Barter, L. Molent, L. Robertson, Using in-service F/A-18 A/B aircraft fatigue cracking as disclosed by teardown to refine fleet life limits, in *Proceedings USAF ASIP Conference*, Jacksonville, Florida, USA, 1–3 December (2009)
19. A.P. Berens, P.W. Hovey, D.A. Skinn, Risk analysis for aging aircraft fleets. Analysis, vol. 1, USAF WL-TR-91-3066, USA (1991)
20. P. White, L. Molent, S. Barter, Interpreting fatigue test results using a probabilistic fracture approach. Int. J. Fatigue **27**(7), 52–767 (2005)
21. Y. Macheret, P. Koehn, Effect of improving load monitoring on aircraft probability of failure, in *Proceedings 2007 IEEE Aerospace Conference*, Big Sky, Montana, USA (2007)

Chapter 8
Fleet Management, Repair and Life-Enhancement Considerations

8.1 Fleet Management Considerations

The following are some lessons learnt from a career in structural integrity fleet management of military aircraft:

1. Safety is paramount but aircraft availability is King.
2. Unanticipated or unscheduled maintenance is to be avoided, since this is very expensive and impacts availability.
3. Avoid grounding the fleet at all costs! Some of the tools presented in this book address this.
4. Do not recommend inspections without well-funded logistic footprints.
5. Align inspection periods with scheduled maintenance: R2, R3, R4.
6. Fleet managers will pay for certainty or insurance to ensure operations late in the life of type.
7. An approximately 10-year period may be required to implement a major modification programme.
8. The manufacturer and maintainer 'seed' your aircraft with flaw-like discontinuities. However, the fleet manager may be able to life the aircraft using the techniques described here-in.
9. Only worry about the possibility of rogue flaws for about a year into service, or late in the life of type.
10. Proper definition is most important, e.g. Prognostics and Health Management (PHM) system without prognostics should generally be called automated NDI.
11. Flaw detection (diagnosis) without prognosis (consequence) is of limited value. The lead crack framework provides one means of prognosis when knowledge of the load history is unavailable.
12. Be the enemy of complexity.
13. There will always be dissenters for new or innovative methods. Learn from them and their objections and criticisms.

L. Molent, *Aircraft Fatigue Management*, SpringerBriefs in Applied Sciences and Technology, https://doi.org/10.1007/978-981-99-7468-9_8

8.2 Modifications and Repairs

Aircraft require highly efficient and durable structures to meet performance and life goals. To achieve this, it is not unusual for manufacturers to rely to some degree on fatigue life-enhancement techniques, i.e. methods to increase the fatigue resistance of local and highly stressed details. The locations for these enhancements may be chosen in the design phase or may be identified during testing or, indeed, during service. It is usual that an FSFT (if conducted) reveals locations prone to fatigue cracking that require remedial action and cannot be easily redesigned, nor can redesigned changes treat some problem locations where unanticipated in-service cracking has occurred.

These enhancement methods fall into two broad classes: (i) those that are aimed at inducing a local beneficial compressive residual stress field and (ii) those that are aimed at reducing the local operational stresses. Some of the techniques available to the fleet manager, should in-service damage or detected cracking occur, are summarized in Table 8.1. The reader is directed to Ref. [1] and Ref. [2] for more details and some pros and cons of each technique.

Three repair techniques that benefit from the LCFLF are summarized in Sects. 8.3–8.5.

8.3 Bonded Repairs

The correct application of bonded composite repairs has been shown to be a highly efficient means of strengthening under-designed locations or significantly retarding the growth of fatigue cracks [3–10]. Bonded repairs are considered superior to conventional metallic repairs for the following reasons [3]:

(1) Metallic repairs are more compliant or flexible than composite repairs and hence may not significantly stop the 'repaired' crack from cyclically opening and closing, thus continuing to grow under service loading. Bonded repairs transfer loads through shear and, given the comparatively large contact area, can stop or significantly retard cracks growing under normal operating loads.

(2) Bonded composite repairs do not introduce new stress raisers in the form of new fastener holes: these could be unintentional new sources of fatigue nucleation and growth.

(3) Unidirectional composites have high directional stiffness and can be tailored to address the principal problem-causing loads without the penalty of parasitic stiffening. Parasitic stiffening attracts loads to the repaired area and can reduce or even eliminate the repair's effectiveness. Notable examples are the less than successful *metallic* patch repairs of Lockheed Martin F-16 bulkheads [11].

(4) Composites have lower density than metals and hence the repairs are lighter and thinner (this is an important consideration for aerodynamic drag resulting from external repairs).

Table 8.1 Some life-enhancement techniques

No	Technique	Brief description
Life enhancement techniques relying on compressive residual stresses		
1.	Shot peening	Small diameter media (e.g. glass, ceramic or metal) with typical diameters from 0.25 to 1 mm are accelerated by a compressed gas to impact a metal's surface and induce deformation and compressive residual stresses that are approximately 0.25–0.5 mm deep
2.	Laser shock peening	A thin liquid layer on a metal's surface is rapidly heated by a high-energy laser pulse, causing it to be vaporized into plasma at temperatures in excess of 10,000 °C. This generates a high-pressure pulse and shock wave in the adjacent metal, inducing deformations and a resulting layer of compressive residual stress, which is typically at least 1 mm deep
3.	Split sleeve cold expansion of holes	Plastic deformation through radial expansion (cold working 3–5% for aluminium alloys) of a hole by sleeving and drawing through an oversized mandrel. This induces a zone of residual compressive stress that may extend radially to about one radius distance around the hole, typically >1 mm deep
4.	Cold expansion of bushings in holes	Cold expanding a bushing that has been placed in a hole via drawing a mandrel through it in a similar manner as the split sleeve process. The bush induces radial compression in the wall of the hole as well as supporting the hole, thereby reducing the stress concentration factor under cyclic loads
5.	Interference fit fasteners and bushings	The fastener insertion/locking process induces radial compression (typically 5%) as well as supporting the hole, i.e. reducing the stress concentration factor under cyclic loads
6.	Shrink fit fasteners or bushings	Similar to interference fit, but achieved by inserting the fastener or bushing *as-cooled* and allowing it to expand on reaching room temperature to achieve the radial compression
7.	Cold burnishing (deep surface rolling)	A high-pressure fluid is used to float a metallic rolling ball in a socket that is hydraulically pressed against the workpiece. The ball rolls freely along a surface to induce plastic deformation and a compressive residual stress layer that is typically at least 1 mm deep

(continued)

Table 8.1 (continued)

No	Technique	Brief description
8.	Cold coining	A hole is straddled by two dies with raised rings (ring thickness may be ≈ 1 mm) that are squeezed to produce an indented ring of compressive residual stresses around the hole

Life enhancement techniques relying on stress peak reduction

No	Technique	Brief description
9.	Reshaping and/or adding material during design	Conventional design practice if any location is suspected to have inadequate fatigue life. Includes machining of shallow radii, thickening local regions (i.e. landings around holes) and introducing doublers as part of the design
10.	Conventional fastened doubler repair	Doublers are simple and often effective at reducing stresses in a component location that is highly stressed or contains cracking, corrosion or mechanical damage. Such doublers are usually fastened in place by bolts or rivets
11.	Adhesively bonded repair	Alternative to conventional fastened doubler that obviates fasteners and the holes that these require
12.	Laser cladding	A form of additive manufacturing where a high-powered laser is used to melt metallic powder (about 50 μm diameter) carried in an inert gas (e.g. argon) to produce a layer/s that welds to a surface
13.	Reshaping critical details by material removal and shape optimization	Mechanical blending is often used to remove damage such as surface cracks and corrosion. The shape of the blend can be optimized using iterative computational techniques to minimize the resulting stress concentration
14.	Stop drilling	A small hole is drilled a few millimetres ahead of a crack. It is a good practice to ream and cold work the hole. It should also be plugged before reinforcing the area with a doubler
15.	Welding	Many types available but only appropriate in specific cases and (very) rarely for the high-strength aluminium alloys used in airframes
16.	Supersonic particle deposition (SPD)	SPD (cold spray) deposits metal by accelerating metal particles to high speed in an inert gas and impacting them on the surface of the part to be built up

(5) Composites have high failure strains but are highly formable so that the repairs can cope with complex geometries.

(6) It may be possible to inspect for cracking 'through' the composite repairs.

Extreme care is required to ensure the mating surfaces to be bonded are clean, and the ends of the repair doublers require proper tapering to reduce peel and shear stresses. The doublers are normally heat-cured under pressure or vacuum with an appropriate adhesive.

Examples of successful repairs include the use of bonded boron/epoxy composite to address cracking in the RAAF's F-111C lower wing pivot fitting [5] and the USAF's C141 lower wing skin weep holes [9]. A failure criterion for the matrix-dominated failure is presented in Ref. [10].

If the composite patch is designed to reduce cracking (rather than stopping it) the resultant FCG will be exponential [8]. If the FCG rate for the unpatched cracked structure is available, then the FCG rate for the bonded repaired can be derived using the cubic rule discussed in Chap. 5: only the stress reduction due to the doubler is required [11]. This allows a simple estimate of continued crack growth by following the LCFLF method. Finally, it was first shown by the author that the stress concentration effect of the ends of the repair can be significant and need special consideration [10]. For example, the repair should never be terminated across a fastener head [10].

8.4 Cold-Working Repairs

Steel shot, glass bead or ceramic bead peening has been used for a long time to improve the fatigue performance of aircraft components [1]. However, the risk of shot peening that covers pre-existing cracks needs special consideration, because such cracks have the potential to invalidate the peening-induced beneficial compressive residual stresses. It is therefore normal practice when blending out a surface-cracked area to perform a 'confidence cut', i.e. extra removal of material after the last NDI that showed no crack indication. Because current NDI thresholds are significant, a confidence cut of at least 1 mm may be required to ensure that no more cracking exists. This can be a significant problem for relatively thin structures.

An alternative procedure has been developed [1] using the lead crack concept. This concept is used to conservatively estimate the likely depth of a crack for a prescribed POF at the time the shot peening is to be done. If this type of problem arises, then the first step in the repair process is to determine the stresses driving the cracking. This can be done by several means including review of the design analyses, strain gauging or finite element analyses. Alternatively, if the cracking on the FSFT or in-service aircraft can be excised and examined using QF, it may be possible to generate a crack growth curve. The driving stress can then be indirectly estimated by (i) comparison with crack growth data generated from fatigue coupon testing of representative material under the test spectrum, (ii) assuming the worst-case FCGR as illustrated in Fig. 7.1 (see Chap. 7) or (iii) in the case of an in-service crack, coupon data generated with the in-service spectrum, conducted at several stress levels as illustrated in [2, 12].

The EPS distribution, e.g. see Sect. 6.2, is used to choose the nucleating flaw size standard deviation (SD) required by the POF. This maximum EPS is then 'grown' according to the LCFLF method to estimate the maximum crack depth at the proposed repair induction time. The (conservative) likely maximum crack depth is the recommended amount of material to be removed. This innovative procedure, although more complicated, generally requires a significantly smaller depth of material to be removed than the traditional confidence-cutting approach.

8.5   Pitting Corrosion Management

Airframe corrosion is a major sustainment cost driver and availability degrader. Despite corrosion prevention or protection schemes/treatments and corrosion prevention and control plans, in-service corrosion does occur and has the potential to impact the structural integrity of aircraft. A significant proportion of aircraft corrosion can be classified as pitting corrosion, which is a form of localized metal corrosion. The pits may be shallow depressions or cavities, and in its early stages pitting may have the appearance of general corrosion at a macroscopic level. Pitting may be present under white or grey powdery deposits of corrosion product on aluminium alloy metal surfaces, and tiny holes or pits are seen after clearing away the deposits. Since pitting corrosion is one form of corrosion that can lead to fatigue cracking, it is specifically addressed here. However, it should be stressed that not all corrosion pitting leads to cracking.

While fatigue nucleating from typical production and surface finish discontinuities has been well dealt with in previous chapters of this book and also, for example, by the US Air Force's Aircraft Structural Integrity Management Plans (ASIMP [13]), the nucleation of fatigue cracks from corrosion pitting has not been addressed much beyond the 'find and fix it' approach. This approach means that when corrosion evidence is found, the only option available is to remove the corrosion as soon as possible and 'recover' the affected area. This generally means an unscheduled repair that can be expensive and impact aircraft availability.

Investigations by Trathen [14], which presented the corrosion rates measured in internal bays in a range of RAAF aircraft, were among the first to report that there appeared to be little corrosion activity occurring during flight. This is supported by the work of many researchers who state that aircraft fatigue cracking occurs in cold and dry in-flight environments where corrosion activity is largely suspended; in other words, corrosion and fatigue are decoupled [15–18].

Here we summarize a basis for allowing detected pitting corrosion to remain in service for a *limited* time, e.g. to the next planned or scheduled servicing. Logistically speaking, unanticipated maintenance costs significantly more than planned and orderly maintenance. By delaying the repair of pitting corrosion until the next scheduled maintenance, it is believed that this will save considerable resources and improve aircraft availability.

The salient points include

a. The area where corrosion is detected has been representatively loaded during the pre-service FSFT for the required number of lifetimes required by the certification basis.
b. A corrosion pit is a fatigue-nucleating source, although the depth of the pit normally exaggerates its EPS; in other words, the pit depth is significantly larger than its EPS [15–18]). In an ideal situation, there will be data available for a corrosion pit EPS distribution, such that a conservative value may be chosen for analyses, e.g. a three standard deviation (SD) pit.
c. There are no known detrimental in-flight environmental effects on possible or actual fatigue cracking.
d. The pitting is detected before fatigue cracking occurs.

The basis of the method is illustrated in Fig. 8.1. Here an FSFT to three lifetimes is assumed and the area of interest has just survived the residual strength criterion (a_{CRIT} = 10 mm). A conservative EPS of 0.2 mm for a 3 SD pit is chosen, and it is assumed to grow at the worst-case (fastest) maximum design rate (i.e. just meets the test requirement without failing). It can be seen that a crack from this 0.2 mm pit would take approximately 1.7 lifetimes to grow to the assumed critical crack depth of 10 mm. Even if the pit were detected late in life (e.g. at 50%) a considerable period of growth beyond a typical NDI detectability threshold of approximately 1 mm still exists (even after the application of a typical safety factor).

In summary, pitting corrosion is known to be a major maintenance driver for aircraft production aluminium alloys (particularly in the peak aged T6XXX condition). FCG (which mainly occurs in the sky) is not significantly affected by exposure to a corrosive environment (which mainly occurs on the ground). Thus, laboratory

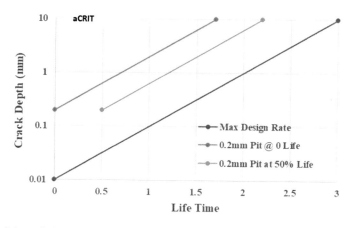

Fig. 8.1 Schematic for the worst-case growth of a 0.2 mm deep corrosion pit (*created by author*)

FCGR crack data generated in air can be used for lifing purposes. It appears for the materials considered in this book, mainly high-strength aluminium alloys, that the effective crack-like size, or EPS, of the corrosion pits is much less than its physical depth. This may not be the case for other materials and should be individually assessed.

The above method provides a safety-based justification for leaving detected and unrepaired in-service pitting corrosion until a planned regular maintenance. The method relies on the *fatigue metrics* of pitting corrosion, namely, the pit's EPS and the lead crack concept. By delaying the corrosion repair until the next scheduled maintenance (rather than the 'find and fix' approach), it is most likely that this will save considerable logistical costs and improve aircraft availability. Therefore, this method is considered to be a major improvement in Aircraft Structural Integrity Management and Sustainment.

References

1. L. Molent, S. Barter, B. Main, Life assessment and repair of fatigue damaged high strength aluminium alloy structure using a peening rework method. Eng. Fail. Anal. **15**, 62–82 (2008)
2. L. Molent, S.A. Barter, Fatigue life enhancement for metallic airframe materials (Chap. 21), in *Aerospace Materials and Material Technologies*, ed. by N. Eswara Prasad, R.J.H. Wanhill, vol. 2. Aerospace Material Technologies. Indian Institute of Metals Series (Springer Science+Business Media Singapore, 2017)
3. R. Jones, L. Molent, A.A. Baker, M.J. Davis, Bonded repair of metallic components: thick sections, in *Analytical and Testing Methodologies for Design with Advanced Materials*, eds. by G.C. Sih et al., North Holland, Amsterdam, (1988)
4. L. Molent, Composite repair of aircraft structures, in *Proceedings AIAA International Aerospace Conference*, Los Angeles, USA, 12–14 February (1991)
5. L. Molent, R.J. Callinan, R. Jones, Structural aspects of the design of an all boron/epoxy reinforcement for the F-111 wing pivot fitting. Compos. Struct.Struct. **11**, 57–83 (1989)
6. R. Jones, L. Molent, Repair of multi-site damage, in *Advances in Bonded Composite Repair of Metallic Aircraft Structure*, eds. by A. Baker, F. Rose, R. Jones, vol. 1, Elsevier Science Ltd., (2002)
7. R. Jones, N. Bridgford, L. Molent, G. Wallace, Bonded repair of multi-site damage, in *Structural Integrity of Aging Airplanes*. ed. by S.N. Atluri, S.G. Sampath, P. Tong, Springer, Berlin/Heidelberg, (1991), pp.199–213
8. R. Jones, S. Barter, L. Molent, S. Pitt, Crack patching: an experimental evaluation of fatigue crack growth. Compos. Struct.Struct. **67**(2), 229–238 (2005)
9. L. Molent, N. Enke, G. Wallace, R. Jones, Thermomechanical analysis of bonded repaired skin splice joint specimens, ARL-TECH-R-24, Defence Science and Technology Group, Melbourne, Australia, (1994)
10. L. Molent, J. Paul, R. Jones, Criteria for matrix dominated failure, ARL-STR-REP-432, Defence Science and Technology Group, Melbourne, Australia, (1988)
11. R. Jones, D. Hui, Analysis, design and assessment of composite repairs to operational aircraft (Chap. 8), in *Aircraft Sustainment and Repair*, eds. by R. Jones, N. Matthews, A.A. Baker, V. Champagne Jr. (Butterworth-Heinemann Press, 2018), pp. 325–456. ISBN 9780081005408
12. R.A. Pell, P.J. Mazeika, L. Molent, The comparison of complex load sequences tested at several stress levels by fractographic examination. Eng. Fail. Anal. **12**(4), 586–603 (2005)
13. MIL-STANDARD-1530D (W/Change-1), Department of Defense Standard Practice: Aircraft Structural Integrity Program (ASIP), October (2016)

14. P. Trathan, Corrosion monitoring systems on military aircraft, in *Proceedings 18th International Conference on Corrosion*, Perth, Australia, 20–24 November (2011)
15. L. Molent, S. Barter, R.J.H. Wanhill, The decoupling of corrosion and fatigue for aircraft service life management, in *Proceedings ICAF Symposium*, Helsinki, 3–5 June (2015)
16. L. Molent, Managing airframe fatigue from corrosion pits – a proposal. Eng. Fract. Mech.Fract. Mech. **137**, 12–25 (2015)
17. S.A. Barter, L. Molent, Fatigue cracking from a corrosion pit in an aircraft bulkhead. Eng. Fail. Anal. **39**, 155–163 (2014)
18. L. Molent, R.J.H. Wanhill, Management of airframe in-service pitting corrosion by decoupling fatigue and environment. Corros. Mater. Degrad. **2**, 493–511 (2021)

Chapter 9
Concluding Remarks

This book summarizes some innovative contributions to the field of aircraft structural integrity, with emphasis on the fatigue of metallic airframes. Examples of the advancement of scientific knowledge include the following:

1. The fatigue-driving mechanisms in constant amplitude (CA) and variable amplitude (VA) loading produce different fracture surface topographies, implying an inherent difference in fatigue behaviour related to these two types of loading. This is one reason why the developed fatigue lifing tools summarized in this book rely mainly on VA crack growth data rather than traditional CA crack growth data.

2. Lead cracks grow in an approximately exponential manner. This behaviour has been integrated into a framework, the Lead Crack Fatigue Lifing Framework (LCFLF), to produce an effective and efficient fatigue lifing model. In particular, this model can be used to assess the significance of cracks detected during service, notably in some cases without knowledge of the applied load history.

3. Confirmation that fatigue nucleates from production-induced and inherent material discontinuities in airframe structures and components. Furthermore, for highly stressed locations the nucleation period is insignificant, i.e. Fatigue Crack Growth (FCG) commences almost immediately when an aircraft enters service or when a Full-Scale Fatigue Test (FSFT) has begun. The significance of the physically short crack growth regime and quantitative fractography (QF) has been emphasized.

4. The typical effective crack-like size (or Equivalent Pre-crack Size (EPS)) of fatigue-nucleating discontinuities in airframes appears generally to be approximately 0.01 mm in depth.

5. Stressing the importance and utility of a representative FSFT. A relatively simple lifing method based on Remarks 2–4 has been used in FSFT interpretations, and also the lifing of aircraft components based on in-service crack findings.

6. The development of several unique FCG prediction tools.

7. Highlighting the importance of knowing the actual load history experienced by individual aircraft to (i) avoid exceeding certification limits during fleet management and (ii) provide a basis for achieving the aims of individual aircraft usage tracking.
8. Proposing that the purpose of an FSFT should be to maximize the amount of applied cyclic load history in order to detect all fatigue-prone locations, rather than the conventional aim of testing to a fixed multiple of the expected lifetime.
9. A means of directly comparing the efficiencies of military aircraft structural design philosophies with the conclusion that the guidance provided by each standard results in airframes with varying tolerance to fatigue cracking.
10. A method of determining the relative severity of different load histories using a single fatigue test coupon.
11. The observation that multiple-site fatigue cracks generally grow independently of each other until the ligament between them fails by plastic collapse, at which point the cracks suddenly link-up.
12. A model that enables blind predictions (i.e. without prior knowledge of results or an initial need to calibrate the model) of fatigue lives of metals subject to VA loadings at various stress levels, using a single crack growth rate curve and a fracture toughness parameter value.
13. Understanding the parameters that govern the scatter in total lives, as observed for nominally identical fatigue tests.
14. A contribution to the probabilistic assessment of failure.
15. The collation and assessment of a large number of crack nucleating discontinuities from actual F/A-18 airframe cracks has allowed probability distributions to be fitted, thus enabling mean discontinuity/defect depths to be established.
16. A number of innovative means of addressing the occurrence and significance of in-service damage, including pitting corrosion.

Many of these contributions and innovations have been implemented successfully in the support of the RAAF fleet and have attracted wider attention internationally. They have resulted in logistical cost savings running into tens of millions of dollars, significant extensions to the service lives of some aircraft types and improved operational availability.